DOWN AND OUT ON
THE FAMILY FARM

VOLUME 15 IN THE SERIES

Our Sustainable Future

Series Editors

Charles A. Francis
University of Nebraska–Lincoln

Cornelia Flora
Iowa State University

Paul Olson
University of Nebraska–Lincoln

DOWN and OUT on the FAMILY FARM

Rural Rehabilitation in the Great Plains, 1929–1945

Michael Johnston Grant

University of Nebraska Press

Lincoln & London

Publication of this book was
assisted by a grant from
The Andrew W. Mellon Foundation.

♾

Library of Congress Cataloging-in-Publication Data
Grant, Michael Johnston.
Down and out on the family farm : rural rehabilitation
in the Great Plains, 1929–1945 /
Michael Johnston Grant.
p. cm. — (Our sustainable future)
Includes bibliographical references and index.
ISBN 0-8032-7105-0 (pbk. : alk. paper)
1. Family farms—Great Plains. 2. Farms, Small—
Government policy—Great Plains. 3. Rural development—
Great Plains. 4. United States—Economic conditions—
1918–1945. I. Title. II. Series.
HD1476.U5 G7 2002
38.1'878'09043—dc21   2002022850

CONTENTS

ILLUSTRATIONS

ACKNOWLEDGMENTS

I have worked both in farming and in scholarship. The two are not as far from each other as one may think. Author Beverly Lowry wrote that "both took long hours and single-mindedness, resiliency in the face of major set-backs, a gift for tedium and a flair for the marriage of the unusual. Both strained the eyes and lower back and depended to some degree on fate, prejudice, perspective, and the intuitive flash." And similar to working with the land, the produce of research is impossible without the assistance of others. First, I would like to thank Carl Strikwerda, Ted Wilson, Lloyd Sponholtz, and Donald Stull who read this manuscript when it was a dissertation. Thanks especially to Donald Worster, who poured through page after page of the drafts and suggested improvements and clarifications. His questions forced me to struggle with the moral significance of farm life in America and the Great Plains. I fondly remember the late Donald McCoy for his generous enthusiasm during the early days of this project. Members of the informal Dissertation Workshop, including Jay Antle, Brian Dirck, Kristine McCusker, Rusty Munhollon, and Chris O'Brien, gave both their support and their valuable advice on earlier drafts. Special thanks and much gratitude goes to James Leiker, who labored through the first drafts of nearly all the chapters. Finally, Deborah Fink generously shared her criticisms and suggestions as this book matured.

Research grants from organizations around the country greatly enhanced my enjoyment of the research and the quality of this book. Part of my research was funded by an Alfred M. Landon Historical Research Grant administered by the Kansas State Historical Society, Inc., Topeka, Kansas; the Hoover Presidential Library Association, West Branch, Iowa; and the Franklin and Eleanor Roosevelt Institute, Hyde Park, New York.

Spending time laboring through the archives of these institutions was one of the highlights of writing this book.

Finally, I would like to dedicate the final product of years of research and writing to my family, both the members living and the deceased. The living members of my family gave me the support, inspiration, and the example of how to start, work through, and complete this difficult process. In closing, I dedicate this work in special memory to my grandparents, Frederick and Florence Johnston and Felix and Melvina Grant, all now deceased. Both couples ran farms and raised families during the 1930s, some of the hardest years in American history. The fact that Johnstons and Grants are still farming in Nebraska is a testament to the strength and perseverance their descendants share.

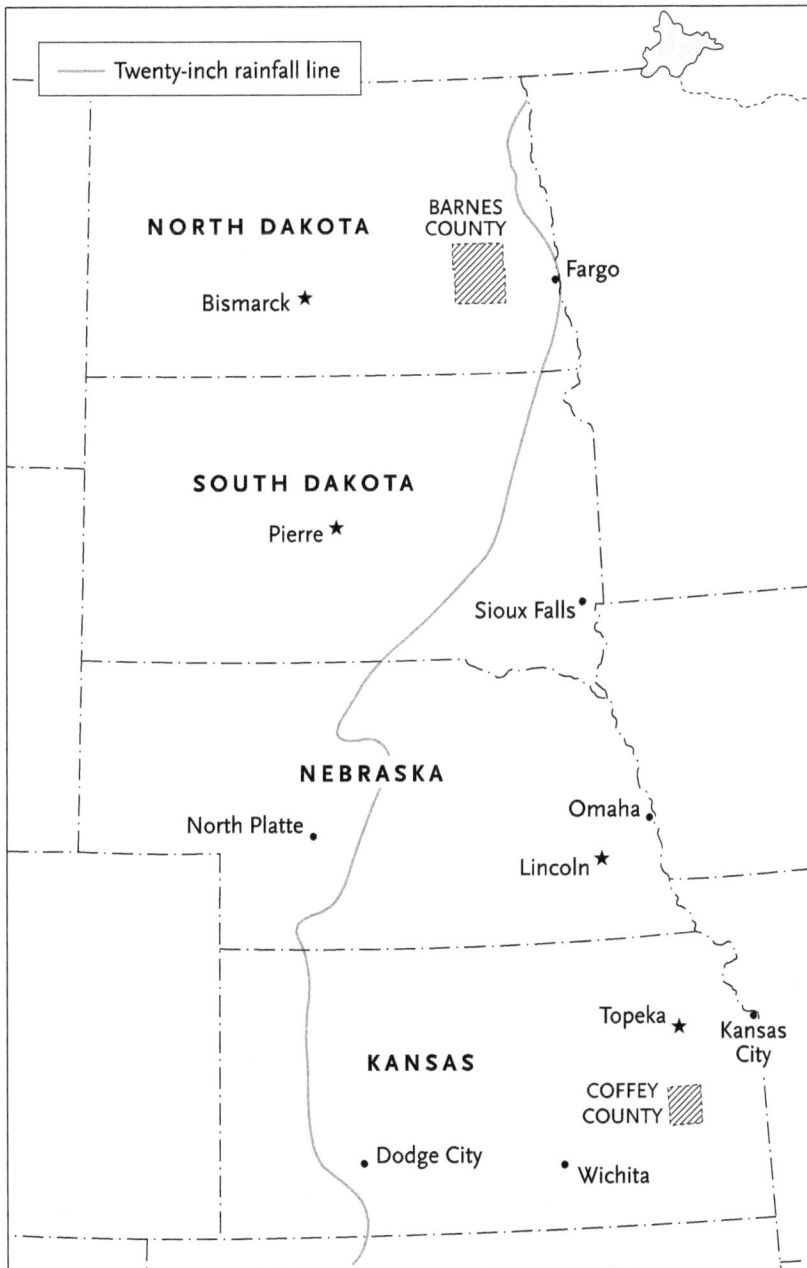

NORTH DAKOTA

BARNES COUNTY

Fargo

Bismarck ★

SOUTH DAKOTA

Pierre ★

Sioux Falls

NEBRASKA

North Platte

Omaha

Lincoln ★

Topeka ★

Kansas City

KANSAS

COFFEY COUNTY

Dodge City

Wichita

The Great Plains, 1929–1945

Introduction

In order to grow bigger, you have to have your neighbor's farm. You have to make
a decision that you would rather have your neighbor's farm than your neighbor.
WENDELL BERRY on American agriculture

Empty farmhouses in the American Great Plains inspire this story. As a
boy growing up in western Nebraska, I rode with my father from our home
in town to our farm thirty miles away. During commutes on cold winter
Saturdays, my father lit up Camel cigarettes (unfiltered) and we talked. One
time, I remember, I looked at the vacated, run-down farmhouses that
dotted the wide-open fields, and I asked my dad why there were no people
in them. This was no easy question, and he thought about it for a while as
the pickup truck jostled us along. "Well," he said, "people got spoiled and
wouldn't put up with living on a farm." Twenty years later, as a historian for
the National Park Service, I visited rural sites in southeastern South Dakota
for a historical survey of the area. One of the sites was the Rising Hail Colony, named after a Native American chief and sponsored by the New Deal's
Indian Relief and Rehabilitation program. Rising Hail Colony was a complex of houses, barns, and a kitchen and recreation hall built of chalkrock
stone from bluffs near the Missouri River. Among the tall marijuana plants
that grew wild around the colony were sturdy homes and work buildings,
now deserted, that once housed and served eight Yankton Sioux families on
the project. They raised their own livestock and crops until 1942.[1]

I was surprised to find a government housing and work project in the
conservative rural plains, already speckled with deserted farm homes. It
stimulated my curiosity about New Deal intervention in the region. This
study of the government's rural rehabilitation program on the plains is the

result of my childhood curiosity and my later visit to Rising Hail. In it I also investigate the clients of this program of the Resettlement Administration (RA) and Farm Security Administration (FSA). Although I was familiar with the FSA through its wealth of photographs of struggling plains farm families in the 1930s, I knew little about the families themselves. Many of them became clients of New Deal programs. I was curious. Just who were these families and what were their farms like?

To answer these questions, this book examines the impact of economic and agricultural changes on a specific economic class of farmers and traces how they reacted to these changes in their livelihood. During the 1920s and 1930s these trends matured and grew more sophisticated. As a result, in regions of the country such as the Great Plains, farmers themselves had to transform their operations to remain on their land.

Most farmers became aware that to advance their standard of living along the lines of urban consumers, they had to use outside capital and tractor-powered farm machinery to cultivate more acres, produce more, and thus expand their income. Just as plants in the region have to spread out their roots to find moisture and survive, so these farm families had to increase their scale of cultivation and production to remain solvent. Unfortunately, the semiarid areas of the plains repeatedly stymied this more expansive scale of farming. This study explores the fate of farmers within the Great Plains who struggled to enlarge their operations to support a middle-income lifestyle. However, many farmers lacked the luck, desire, managerial skills, capital, and land to fully take advantage of modern, mechanized agriculture.

Plains farmers were among the first to convert to capital-intensive, large-scale agriculture. However, their problems were not inherent in the region or in farming alone. Particularly after World War II, American farmers learned to either get bigger or get out. Twentieth-century agriculture was distinct because productivity increased rapidly in commodities and in land and labor. Like other industries, working the land in the modern world means producing more with greater efficiency—or abandoning the game.

This study is set in the years 1929–1945 in the plains states of Kansas, Nebraska, North Dakota, and South Dakota. It focuses on the role of the federal government in both supporting and correcting the domestic and agricultural practices of rural families. Called "rural rehabilitation," it consisted of government-supervised loans and grants to farm families combined with technical advice to reform life in the field, farmyard, and farm home. The period between the Wall Street crash and the end of World War II was a time when agricultural, economic, and environmental trends thwarted farm families making the jump into the middle-income level. The rising farming costs of post–World War I America and higher standards of living collided with the disastrous drought and economic depression of the 1930s. These forces placed many aspiring farmers on the borderline between poverty and economic security. I call this group the "borderline farmers," and they are the subjects of this study. I also examine regional support for and opposition to government attempts to rehabilitate and reform these farmers' practices.

There are two definitions of the setting, the Great Plains. For the purposes of this study, the plains is a political region made up of the states of Kansas, Nebraska, and North and South Dakota. Geographically, the plains is an area west of the Hundredth Meridian that normally has twenty inches of rainfall or less annually. While farmers in the western semiarid Wheat Belt and the more humid eastern Corn Belt of the plains states had different environments to contend with, their situations and problems are similar enough to group them together for this study.

This book does not incorporate the "buffalo commons" argument into its pages. This argument contends that the semiarid areas of the western Great Plains never should have been settled by European-Americans. Champions of the buffalo commons see modern plains agriculture as an environmental disaster destroying the ecosystem. They assert that only expensive federal subsidies for farmers and the infrastructure, rapacious irrigation, and the ruinous use of fertilizers have floated the region and kept "America's outback" from bankruptcy. Their solution is to turn the region into a natural preserve where bison and antelope, not humans, are dominant.

These critics have performed a great service to the Plains region. In the rush to dispose of public lands quickly, the federal government did indeed sell or give lands to individuals unable or unwilling to serve as stewards of the land. Critics of American land policies properly question the moral basis for farming practices that were blindly copied from the more humid Midwest to the dryer western reaches of the plains. They note that many counties in the western plains support fewer than six people per square mile and are, in fact, still frontier counties. These critics demand regional cooperative planning where base individualism has dominated, often with destructive results for the land. European-Americans, they correctly observe, fraudulently and violently displaced Native Americans from the region, often to harm the environment with agricultural practices that cannot be sustained indefinitely.

However, the American Great Plains is not a desert. The western plains region certainly is a deceptively fragile environment. Even long-term farmers there sometimes question whether the area, with its isolation from markets, harsh droughts, high winds, and monumental temperature extremes can support agriculture. Still, in any landscape it takes generations to learn how to till the soil and keep livestock on pastures. One of the lessons learned is that some of the region's land never should have been brought into production. But this is true not only in the plains but in America as a whole. During the 1930s the federal government correctly purchased and withdrew millions of farmed acres from production and put them in land utilization projects. While most of this land was in the dry parts of the West, the government also took farmland in distinctively humid areas such as Wisconsin and Florida and converted them into national grasslands and national forests.

Furthermore, since World War II nearly all agriculture has required inputs such as fertilizers, government subsidies, and expensive infrastructure. Plains agriculture made the conversion to this type of farming earlier and is simply more unstable than most regions. It is my contention that much of the plains can support farming and ranching. However, the small to mid-sized farms and their families that are the topic of this work could not adjust to this new, expensive, large-scale form of agriculture. It is this larger scale of farming and the resulting problems that bind the western and eastern plains together. While in most of the plains large-scale agriculture is viable, it was beyond the means of many of its farmers, as this book will show.

It is essential to detail what this book is and what it is not. First of all, it proceeds from Sidney Baldwin's *Poverty and Politics: The Rise and Decline of the Farm Security Administration*. Baldwin looked at the FSA from a national perspective and focused primarily on the political background of the agency and its various programs. These activities ranged, among other things, from planned resettlement projects and rural health agencies to its famous photographic section. My book, on the other hand, focuses on one of the FSA programs, rural rehabilitation, in the plains region. It also explores the agricultural conditions and political support and opposition for rural rehabilitation in the plains.

This is a quantitative investigation of farm life. Overall, the evidence is economic and political. It explores the interaction of government, economic and political values, and environmental factors. A driving force of much of the action is the growing scale of agriculture. This study explores a very human endeavor—family economics. One of the most important times of the year for Americans is when they sit at the kitchen table and review their finances. Historians tend to ignore the more mundane stuff, but bills, bank accounts, and pay stubs set the limits of our hopes for the future.

Like all investigations, this one had to start with an assumption. I assume that nearly all plains farmers in the 1930s were commercially oriented, that is, the market was the primary road to opportunity to them. Traditionally, opportunity in American agriculture came from two sources: First, the speculative gain on higher land values, and second, the sale of agricultural produce beyond that used for subsistence. As plains farmers harnessed the two in order to purchase consumer goods they desired, they acquired values resembling those of the small-town businessmen and women. Even though the Great Depression forced many to question the capitalist values that contributed to their mess, few deserted commercial farming and its attendant values to embrace subsistence farming. At the beginning of the 1930s, only 2 percent of plains farmers were "self-sufficing," according to the USDA. Although the percentage of plains farmers who consumed most of their produce grew to 16 percent by the beginning of World War II, they were probably driven to this by poor commodity prices. At the end of war prosperity, only 4 percent of plains farmers remained in this nonmarket class.[2] Plains farmers were in fact using business principles to maintain and advance their rural way of life. Even plains religious groups like the Mennonite farmers, who otherwise eschewed modern life, took part in the market economy to preserve their faith, culture, and social order.

That farm men and women, especially those on the borderline of poverty, retained their capitalistic values may seem paradoxical. But westerners have historically preferred opportunity to security. The American West since settlement by European-Americans has been a land of vast wealth combined with social, economic, and environmental volatility. This explains both the region's hardships and its appeal for its inhabitants. The plains of the West deceptively offered enormous opportunities to prospective landowners at low initial costs. However, unstable market and environmental conditions meant migration to the region amounted to a high-odds gamble for farmers.

Most plains farmers saw themselves as capitalistic entrepreneurs engaged in commercial agriculture and operating as individuals, even though their actions often contradicted this. Ideologically, they were capitalistic since they defended private investment and ownership. Farmers hoped to profit from a marketplace that set prices and production and facilitated the distribution of their goods. They were entrepreneurs because they saw themselves as running their own businesses and taking their own risks, again for the sake of profit, which was, after all, the goal of commercial agriculture. Finally, farm men and women were intent to farm as they pleased. They praised their own "rugged individualism" which gave them the freedom to succeed or fail on their land by their own initiative.

Each of these concepts supported the other, yet farmers both embraced and ignored them in the 1930s. This book shows how, despite their ideology, plains farmers allowed and even demanded that government regulate their land use and remake the marketplace. Under the New Deal, the federal government assumed much of the risk in farming, in order to profit farmers. Finally, through voluntary programs these "rugged individualists" surrendered part of their autonomy to the government because of the disastrous economic and environmental climate of the early to mid-1930s. This, of course, contradicted their ideals. By the latter 1930s many people in the plains accepted New Deal commodity and credit assistance as their entitlements from the government. However, they rejected programs they perceived as limiting their autonomy and opportunity or that seemed to propel New Deal reforms further into the countryside. The rural rehabilitation program died as a result.

In her *Main Street in Crisis: The Great Depression and the Old Middle Class on the Northern Plains*, Catherine McNicol Stock addresses the contradictions in the region between individual entrepreneurs and the New Deal they were forced to accept. She cites how the old capitalistic "moral economy" of rural and small-town middle-class plains residents benefited from, but still resented, the mass welfare state of the New Deal. Stock describes them as "petty producers," farmers and small-business owners. Despite the activism of groups like the Farmer's Holiday Association, I argue that most plains farmers accepted, with some reservations, the tenets behind commercial agriculture.

Plains farmers dealt with the importance of economic scale, just as urban and small-town society did during the thirties. Even during the economic doldrums of the Great Depression, businesses across the nation had to grow in size or fail. Increasingly, larger enterprises learned to take advantage of the economies of scale to lower costs in retail, distribution, production, and processing networks to succeed where smaller businesses floundered. Across the country, the large-scale corporation became the model business. American agriculture also advanced in the scale of operations. My subject, however, is not the model farm family. Rather, I will examine the borderline farmers who were out of synch and overwhelmed by the expansive scale of farming. Those in the semiarid western regions of the plains were under added pressure as they attempted to operate in continual and severe drought conditions. Many were able to take advantage of the area's low costs of farming per acre. Many were not.

My focus is on work, not identity. People make their living from what they do, not who they are. It seems so much of today's historical writings compartmentalize people in the past by their identity. My subject farmers are defined by their income level, which determined many of their agricultural practices and opportunities. This study does not use class, gender, ethnicity, and race as categories of analysis. I simply did not have the evidence to determine farming practices during the 1930s and World War II based on these factors.

Class is too a vague a term to use in the rural plains during the 1930s. Surely, people at the time treated each other differently based on their perceptions of class. However, my evidence wouldn't allow me to label families by class. I use the terms lower, working, middle, and upper classes as mostly cultural terms to identify broad sectors, but not to apply to particular farm families in this study.

Men and women of this generation experienced farm life differently and looked at it differently. I did not presume that there was an "internal unity" of farm couples on how they approached their livelihoods. However, I did not investigate how farm men and women looked at their farms, but rather at how they operated their farms and what they produced. I tried very hard to give equal voice to men and women in my evidence. When there was an obvious distinction between women and men, such as the rural rehabilitation program's separate supervision of farm women, I spotlight it.

Similarly, I don't assume that all ethnic and racial groups in the plains region farmed the same way. First-generation Norwegian wheat growers in South Dakota naturally looked at their operations differently than third-generation native-born Nebraskans in the Corn Belt. However, this work doesn't analyze in depth farming practices based on ethnic background. Furthermore, other than a smattering of African American and American Indian farmers, the farmers discussed are white. Finally, the overwhelming majority of heads of these farms were male. Studies based on class, gender, ethnicity, and race distinctions in rural America are essential to fully understanding this nation's history because they reveal the complexity and richness of farm life in the plains. However, income levels influenced agricultural practices and opportunities more than the farmers' backgrounds, acreage, or whether the land was owned or rented. That is why I used income as my yardstick. Accordingly, this book benefited from the wealth of published census and state and federal agricultural data on farm families in terms of annual income.

Especially after World War II larger farms in the plains absorbed the smaller ones. Economic and agricultural trends of the 1930s set the stage for this as small- and medium-sized operations battled to stay above water. When the rural rehabilitation safety net was removed, these smaller operations often failed. It would be easy to focus on a "culprit" to blame—some group of rich landowners or haughty government officials— but that would be unrealistic. The actual culprits are the farmers, the environment, and the markets, and the way they played off each other during the 1930s. General trends during this time played against smaller farmers, whose expectations of the soil were often too high. The Great Plains seems to attract high-stakes gamblers trying to make a living or grow wealthy despite too little investment and a volatile land. Unsubstantial farmers were trying to make it in a larger farmer's world.

Were these larger farmers part of an effort to drive their smaller neighbors off the land? No. They did not drive them off the land, but eventually they opposed New Deal efforts to prop up smaller farms. The Farm Bureau, which represented more prosperous farmers, helped destroy rural rehabilitation during World War II because its members wanted the land and labor of smaller farmers. Rather than culprits, they were simply acting as businessmen and women. In the plains, larger farmers adapted to the increasing scale of agriculture and wanted to grow even larger to keep up with trends. If there is a culprit in this story, it is the proclivity of Americans like the plains farmers to favor opportunity over their own security.

From a long-term perspective, probably no government program or economic system could make small-scale agriculture viable in the plains. Still, I felt some sympathy for these "marginal" farm families and their fight to keep their farms afloat, and it crept into my investigation. It was difficult to write an entire book about them without admiring their tenacity and courage against the odds. Most Americans have the same sympathy when they see a small, locally owned shop struggle to succeed when a huge discount chain store opens nearby. Is there a place for small-scale family farming in the plains? Not likely, but their struggle is worth telling and their passing is worth mourning.

This analysis is important and worthwhile for three reasons. First, it helps explain the most important global migration trend of the twentieth century: the movement of rural folk into towns and cities. In America and around the world millions of families cut loose from their agrarian roots for a life of more opportunity and better living conditions in towns and cities. Between 1929 and 1945 the American farm population fell from 30.5 million to 24.4 million. Farm dwellers fell from 25 to 18 percent of the total population between the Wall Street crash and the end of World War II.[3] This trend also removed much of the vitality (as well as the congestion) of the world's rural areas. According to the United Nations, the world's farm population remained virtually the same, 1.3 billion people, between 1937 and 1950. However, relative to the general global population, the proportion of the farm population dropped from 62 to 55 percent by 1950.[4]

Second, this analysis is significant because the trials of plains borderline farmers during the 1930s reveal the tension in American economic values which exist to the present. Throughout this nation's history Americans have celebrated the moral, political, and social benefits of a large class of small entrepreneurs. This country has praised the "petty bourgeoisie," from the yeoman farmers, shopkeepers, and homesteaders of the past to the family farmer and today's small entrepreneur. However, the forces of the equally celebrated free-market economy are often in conflict with the

interests of small-business men and women. This study of the plains borderline farmer and the rural rehabilitation program shows how far this nation worked to protect a revered class of families, and at what point the
country released them to rough economic currents of the times.

Third, this study is relevant because it addresses the economic and political legacies of the 1930s. The Great Depression remains the paradigm
for comprehensive economic failure in the United States. Whenever Wall
Street becomes skittish, unemployment rates spurt upward, or a rash of
bank failures makes the news, people invariably look to the Great Depression as the cautionary decade. Furthermore, Americans are still struggling
to come to terms with the legacy of the New Deal that addressed the problems of economic crises. Whenever the merits of the "welfare state" or federal intervention into private life are debated, the press and academics
summon the ghost of Franklin D. Roosevelt and his New Deal programs to
explain why the growth of the federal government was necessary in the
first place.

So, as the story begins, settlers migrating to the Great Plains after the
Civil War saw a region filled with promise and opportunities. Homesteaders moved to the plains to take advantage of cheap lands and new transportation networks to become secure and prosperous landholders. But
these pioneers found they had to adjust to a new environment. Because of
cheaper land, more expensive labor, and the diminished rainfall compared
to the humid East, plains homesteaders attempted to acquire and cultivate
as much land as possible as cheaply as possible. This extensive rather than
intensive form of agriculture required larger fields and new, horse-driven
machinery. Furthermore, because of the high costs of converting the
plains into productive commercial operations, many pioneers became en-
meshed in the region's mortgage system. Thus was established a heritage
of financial insolvency in the plains during times of drought and economic
slumps. While many settlers hoped to become successful commercial
farmers, they fell victim to volatile agricultural conditions and unstable
farm income and acquired high debt loads. In response, during the 1870s
and 1890s many of these distressed farmers joined insurgent agrarian
protest groups such as the Populists.

By 1920 plains residents celebrated their prosperity. Agricultural demand during World War I caused an explosion in both commodity prices
and land values. This booming farm economy allowed plains farms to

modernize with more tractors, automobiles, and telephones than any other region in the country. However, the war prosperity was deceptive. Foreign commodity markets were shaky and collapsed after 1921. Furthermore, plains farmers contended with higher operating, labor, and transportation costs that cut their profits. Between the mid-1920s and the Great Depression, however, the price of farm goods made a wobbly upward climb.

Therefore, the decade of the twenties was anything but roaring for many plains farmers. Many of them had overexpanded their debt and productive capacity during the decade. Furthermore, their expectations had also ballooned so that they anticipated a standard of living in parity with the urban middle class. Unfortunately, both plummeting land values and anemic commodity prices deprived them of this. To bridge the gap between their lower income and their higher expectations, many farmers took out second mortgages on their land and became indebted to local merchants. Plains farmers had used commercial credit since the pioneer days. However, farm loans in the region were often risky and difficult to repay. As a result, farmers seemed to fall behind America's cities in wealth and modernization. The plains farm sector with its increasing grain surpluses, high farm mortgage debt, and falling land values was ill-prepared to face the coming catastrophe of the Great Depression beginning in 1929.

Borderline Farmers

Marginalized Farmers on Marginal Farms

In November 1933 Lorena Hickok, a roving relief investigator and confidant of Eleanor Roosevelt, visited Bottineau County, North Dakota. She was shocked to find such profound destitution among many farmers. Farm residents of the county along the American-Canadian border, Hickok reported, were short of bedding, clothing, and fuel. For instance, at the farm of one of the "better off" families on relief, she found two young boys running about "without a stitch on save some ragged overalls. No stockings or shoes," she reported. "Their feet were purple with cold." The kitchen floor was patched with scraps from a boiler and old car license plates. Plaster was coming off the walls.[1] While visiting southwestern North Dakota one month earlier, Hickok observed many farmers who, under normal circumstances, would be in no way considered indigent. Most of the farmers were considered "well-to-do" with 640 acres and thirty to forty head of cattle, a dozen or more horses, and a few sheep and hogs each. Hickok reported that their livestock were "thin and rangy . . . trying to find a few mouthsful [sic] of food on land so bare that the winds pick up the top soil and blow it about like sand. . . . Their [milk] cows have gone dry for lack of food. Their hens are not laying. Most of their livestock will die this winter. And their livestock and their land are in most cases mortgaged up to the very limit." She added that the farmers were "way behind on their taxes, of course. Some of them five years!"[2]

The Great Plains states of Kansas, Nebraska, and the Dakotas offered relief investigators similar scenes of rural poverty in the mid-1930s. Hickok's report covered just a small region where rural families were descending into poverty. Many of these farmers had the land, livestock, and

potential for bountiful yields, but they lacked security. Only five years earlier, the rural plains would have impressed the casual observer. In the late
1920s agricultural production was up, and plains farmers were using tractors and buying Fords and Chevrolets at a rate that put the rest of the country to shame. Five years later the hopes of the Roaring Twenties and agricultural prosperity seemed hollow to farmers in Bottineau County and
elsewhere on the plains.

Indigence was no stranger to rural America, even before the Great Depression. But normally the impoverished were new settlers or were those
with limited resources. The families Hickok visited were established ones
who held property such as livestock, farm machinery, and at least one automobile. They either owned or rented a large amount of land, at least by
national standards. Because they held property, these families were by no
means paupers. How could they be just skirting destitution? The answer
lies in the collision of long-term agricultural trends, environmental problems, and markets flooded with farm commodities, all during the 1930s.
Farmers and farm experts suggested different strategies to stay afloat. Still,
many plains farm families fell into poverty. One set of people, the borderline farmers, had unique problems during the depression. This chapter explores the factors mentioned above and their impact on borderline farmers. Later chapters will review how the New Deal's rural rehabilitation
program addressed the borderline farmers' problems.

The Price Tag on Large-Scale Production

Even without the drought and low market prices, long-term agricultural
trends would have shaken the rural plains in the 1930s. These trends were
the rise in farm productivity, increased farm mechanization, and the increased need for capital investment. The 1930s were a time of relatively
small spurts in productivity per acre in the United States. When measured
in terms of labor and technological inputs, the productive capacity of the
American Great Plains showed only incremental gains. However, what
plains farmers lacked in productivity per acre, they compensated for with
larger farms and opening new land to cultivation. In the four plains states,
the size of the average homestead increased from 390 acres in 1930 to 439
acres in 1940 to 496 acres in 1945, a total growth of 22 percent in fifteen
years.[3] Furthermore, during the 1930s, despite the adverse agricultural
conditions, the total amount of land in farms grew 5 percent in the plains.

More land in production contributed to the larger grain surpluses and lower grain prices. During the late twenties and early thirties, the plains were awash with wheat and corn. Lower worldwide demand for grain combined with a series of bumper crops drove the price of a bushel of wheat down from $1.05 in 1929 to 38 cents in 1932. Corn fell from 77 to 32 cents a bushel during the same period.[4] After a nadir in the grain markets in 1932, the price of wheat and corn rose sporadically during the 1930s, due partly to New Deal price supports. Unfortunately, drought and insect pests such as grasshoppers cut plains farmers' yields, and as a result they were unable to take advantage of higher farm prices.

A second trend, that of technical applications and mechanization, greatly affected the plains. The 1930s saw a blossoming of new technologies in the countryside in seed and machinery that combined to increase agricultural output. Newly developed grain seed increased productivity when USDA scientists created the Marquis strain of hard-red spring wheat resistant to the cold of the northern plains winters. They also used the Turkey strain in crossbreeding to enhance disease resistance, early ripening traits, and quality in wheat. In addition, hybrid corn seed adapted to the soils and climate of the Midwest reached the market. This seed was heavy yielding and amenable to mechanical harvesting.

Also during the 1930s in the plains, tractors continued to replace horses for field work. Yet a shrewd farmer in the 1930s, especially one with less income, thought carefully before embracing tractor power. On the favorable side, small-grain crop farming on the mostly level plains was well suited for mechanization. Beginning in the 1920s International Harvester, Ford, and John Deere introduced tractors for farms. These machines pulled implements such as cultivators and plows, and drove auxiliary cutters and threshers during harvest.

On the other hand, farmers saw several drawbacks to converting from time-tested horses and mules to relatively expensive equipment. In 1930, except for the growing of enhanced strains of small grains such as wheat, adopting power machinery appeared of questionable value. Converting to tractor farming meant having to pay fuel and repair bills for the first time, as well as purchasing implements specifically suited for power farming. Some maintained that tractor farming during the 1930s was actually a losing investment, since the climate and the sometimes anemic commodity markets canceled any profits. Furthermore, using a tractor in the semiarid areas was simply not profitable for farms smaller than 360 acres, regardless of the economy.

Despite the reasons *not* to mechanize, the adaptation of new technology in the American Great Plains had a decisive impact—but not on farm productivity per acre. Most important, new technology cut the amount of labor needed to cultivate a given acre of land. Many horse-drawn machines required an additional worker to operate the attached implement. Tractor-pulled farm implements didn't and ran faster and longer. Tractor farmers had impressive results. The number of man-hours required to grow and harvest an acre of wheat declined from 10.5 hours in the late 1920s to 7.5 hours during World War II.[5] This freed the farmer from having to feed and provide for horses, mules—and hired hands. As a result, by 1940, 55 percent of plains farms, compared with 23 percent of all American farms, had a tractor.[6]

Farm mechanization and the increased scale of farming had different consequences on operations of different sizes. Significantly, it squeezed out the smaller, aspiring farmer who desired the savings that running a larger mechanized enterprise afforded. Between 1920 and 1945 in the plains, the number of farms with more than one thousand acres and those with fewer than fifty acres increased. On the other hand, the number of operations between one hundred and five hundred acres decreased.[7]

Mechanization allowed plains farmers to cut back on labor costs and enabled them to run larger operations. By the mid-1930s the lure of the tractor for many larger farmers was irresistible. Higher commodity prices by that point meant substantial farmers could afford the all-purpose Farmall tractor. Engineering improvements had made the tractor more reliable and less expensive to operate. Plains farmers viewed tractors as a means of increasing the size of their operations while cutting labor costs, thus bolstering profits. Even though running a tractor did not always reduce operating costs per acre, it allowed owners to cultivate more land and earn greater income. Despite the expenses of mechanizing, once converted to power farming, farmers rarely returned to the horse. They gave economic and personal reasons. Despite the romantic image of a horse-drawn plow, working with draft animals was often filthy business. As one farmer reminisced, the operator who rode behind horse-drawn machinery was "in the direct line of fire of all discharged solids, liquids, and gasses, usually disposed of in enormous quantities."[8] This helps explain why, in spite of the

economic hardships of the Great Depression, farmers stuck by their tractors. Val Kuska, colonization agent for the Burlington Railroad, noted,
"The surest way to drive [farmers] . . . from the farm is to take away their
power machinery." One man told Kuska, "'My wife will not run a hotel for
hired men and I'll not run a livery stable on the farm.'"9

This leads to the third agricultural trend—increased capital investment.
The 1930s witnessed continued growth in farm investment, begun during
World War I. However, the additional acreage and machinery had a price
tag. The vast amount of debt owed on the property, livestock, equipment,
and for the payment of operating and living expenses was a primary curse
for the 1920s and 1930s plains farmer. By 1940 three-fifths of the region's
farm owners had mortgaged their property.10 Faced also with anemic commodity prices and pitiful harvests during the drought years, they found it
difficult to stay afloat. The result was that many farmers passed from insolvency during the 1920s to insolvency and poverty during the 1930s.

Throughout the thirties individual farm families faced the mammoth
task of converting their debt load into some form of economic security.
During that decade many realized that no matter how hard they worked,
drought and low productivity only eroded the income needed to repay their
liabilities. The farmer's growing dependence on outside capital to make
the enterprise profitable was both symptom and cause of this income erosion. Farmers walked a desperate tightrope of changing rainfall, fluctuat-
ing commodity prices, interest rates, and transportation costs. Any one of
these could transform their business from a barely profitable operation
into a failure.

Farmers of this era controlled production expenses by substituting human labor with machines and by increasing productivity in the land and
livestock. They accomplished this through higher capital costs. Many farmers of the American Great Plains during and after World War I gambled on
continued high commodity prices, and they borrowed to buy land. Unfortunately, with the fall in commodity prices in the early 1920s, land values
plummeted along with dreams of wealth. Much of the countryside was burdened with debts remaining from the 1920s, a situation exacerbated by the
Great Depression. Since farmers could not profit from rising land values
during the devastating economic depression, they had to depend on their
productive capacity to make themselves profitable. Those farmers with the
means did so by mechanizing and acquiring even more land.

However, those who failed to make the conversion contributed to the Great Plains' high foreclosure rate, which led the nation in the 1930s. The nadir was 1933, when there was an average sixty-two foreclosures per thousand farms in the four states. That was well above the national rate of thirty-nine foreclosures per thousand.[11] Within the plains region, North and South Dakota had the worst foreclosure records between 1926 and 1940, when nearly seven thousand Dakota farmers were forced off their land annually.[12]

These foreclosures made the Great Plains a leading center for "distress transfers": foreclosures and assignments of land to avoid foreclosure. Between 1925 and 1939 South Dakota, North Dakota, Nebraska, and Kansas placed first, second, third, and ninth in the number of average annual distress transfers in the nation.[13] Not until the end of World War II did the specter of farm foreclosures fade in the plains.

Examining farm debt at the county level offers an understanding of the problems plains farmers faced. A 1930 study of farm debt in Potter County in central South Dakota found farmers depended increasingly on outside sources for credit. In 1890 55 percent of Potter County farm owners operated without mortgage indebtedness; by 1930 only 27 percent were without mortgage debt. In that year, the average Potter County farm debt was $8,075. Real estate mortgage liabilities accounted for 72 percent of the total debt, livestock was 13 percent, and miscellaneous liabilities constituted the rest.[14]

The prevalence and the amount of debt were troubling. That few farmers paid off their liabilities, even before the full force of the Great Depression set in, was alarming. The average Potter County farm made $4,383 in cash from its operations in 1930. But after Potter County farmers subtracted expenses (including $479 in interest payments), they had only a little more than $100 to apply to their financial obligations. Farmers satisfied their creditors by taking on an added $232 in added debt in 1930. Farmers were only digging themselves deeper into a pit of red ink.[15]

Apprehension spread over debt-ridden Dakota farmers. H. S. Ewen of Carrington, North Dakota, informed Democratic presidential nominee Franklin Roosevelt that he was selling off his assets, including land and bank bonds, to pay his taxes and insurance premiums. This was still not enough to cover his liabilities, though. To make matters worse, Ewen claimed, twelve of his county's banks had gone bankrupt since 1920, taking with them their depositors' money. Looking over the rural situation, Mr. Ewen advised Governor Roosevelt, "[Y]ou sure have got an awful mess to clean up until this farm problem is solved—[but] it never will be cleaned up."[16]

Exacerbating insolvency in the rural plains was a credit system ill-suited for agriculture. Ideally, during times of falling income, farmers fell back on loans with low rates and a long-term repayment schedule, allowing them to stay afloat until prosperity returned. During the early 1930s most plains farmers had to depend upon five- to ten-year mortgages paid with 5 to 7 percent interest rates, not exorbitant terms. However, until the mid-1930s the main credit sources were land banks, life insurance companies, commercial banks, individuals, and businesses providing credit for farm customers. Under accepted business principles, these lenders denied chancy loans for farmers already in arrears. For example, in 1936 the cashier for a bank in Valley, Nebraska, reported to the governor that seven local farmers owing an average of $1,500 were in danger of foreclosure. He admitted that "under the present set up for a commercial bank there is not any place that these men can be accommodated through the regular channels." Yet he pleaded that "all these men need is time and fair crops. There is not one of them that will not come through if they can have a chance and it seems a tragedy that they are facing a foreclosure."[17] Primarily under the influence of the federal land bank system that arose after 1932, mortgage repayment schedules were lengthened up to forty years.

Outside credit sources added extra costs to farmers' bulging debt load. During the 1920s mortgage and insurance companies purchased many farm loans when the improving farm economy of the late 1920s made land a good investment. Some observers complained that this arrangement placed borrowers at the mercy of greedy company agents. One claimed that when farmers tried to renew their loans, loan companies assessed fees of up to $75 per $1,000 owed, in "Al Capone fashion, trying to exact an exorbitant commission."[18]

The costs of land, machinery, and credit were obvious to all. Even those new to the United States and the Great Plains assimilated the new agricultural trends. The plains states, particularly the Dakotas, had some of the highest rates of foreign-born farmers in the nation before World War II. The region was rich with immigrants from northern, central, and eastern Europe. Some historians contend that foreign-born farmers and their children ran their farms differently from native-born farmers. But it is misleading to bunch these groups into a single category of "foreign-born" farmers. Plains farmers with Norwegian, German, and Czech backgrounds, for example, varied in politics, values, religion, and other important attributes. That said, it appears that by the 1930s the farming practices of foreign-born farmers were similar to their native-born counterparts. A cursory survey of four plains counties with large proportions of foreign-born farmers, as well as those of foreign-born parentage, gives a brief portrait. The following counties and their dominant ethnic groups each bordered counties with mainly native-born farmers in 1930 and can be compared for agricultural traits: Divide County, North Dakota (Norwegian); McPherson County, South Dakota (Germans from Russia); Saline County, Nebraska (Czech); and Ellis County, Kansas (Germans from Russia).

By certain criteria, these foreign-born farmers and their children remained distinct socially and in the way they operated their farms. They had slightly larger families, were somewhat less frequently mortgaged, and tended to use more of their farm produce for their own needs than did native-born farmers. But in critical aspects, counties populated by immigrant farmers were much like those dominated by native-born farmers. Foreign-born farmers owned rather than rented their lands at roughly the same rate, ran operations of similar value and size, and had a similar gross income compared to native-born ones.[19] Other than these categories from the 1930 Census, there exists little published data grouped by origin of the farm operator. It is beyond the scope of this study to investigate the precise agricultural practices of first- and second-generation plains farmers, but available evidence suggests that they kept down costs as much as possible while following large-scale agricultural practices in the plains. Thus it appears foreign-born farmers and their children attempted to be more self-sufficient in their labor, credit, and subsistence needs than native-born ones. But they used these means to operate farms equal to the size, value, income, and ownership rates of native-born farmers in neighboring counties.

Immigrant families had distinctive cultural attributes seen in their religion, food, language, and education. Cultural factors did influence critical aspects in agriculture such as residential clustering, persistence rates, and land inheritance strategies on the frontier.[20] However, their impact on farm practices during the 1930s is harder to gauge. It appears immigrant farmers were more like than unlike native-born plains farmers during the 1930s.

The triad of increased acreage, mechanization, and the need for capital was both blessing and curse to all plains farmers, regardless of birth. Through these trends farm families made a larger income, and standards of work and living comparable to those of the urban middle class. However, during the depths of the Great Depression, when the rural economy collapsed around them, many farmers found fewer advantages to more land, machinery, and outside capital. After assuming intractable debt to purchase land, machinery, appliances, and consumer goods, the rural plains dwellers found themselves unable to pay for them. Farmers were caught in a "cost-price squeeze" as the cost of modern agriculture outran their income. Clearly, it was impossible for many small and medium-sized operations to survive during the 1930s. Even without the depression, the increasing scale of farming would have driven many of them out of business. Economically, there were simply too many smaller farms trying to make it in a changing world. The depression drove even larger farms out of business. Under normal conditions, many struggling farm families (and the American countryside) would have been better off if they had sold out and moved into towns or cities. That option was closed in the face of economic depression that devastated industrial America.

Rooms Full of Dust and Wings of Yellow Cellophane

Decreased commodity prices and high debt levels vexed plains farmers, but they felt truly damned by two other trends. A series of droughts and insect plagues hit all sections of the American Great Plains during the 1930s and caused ruinous crop failures. Of the two, the lack of moisture had the greatest impact. The western plains historically received an average of twenty inches of rain annually. In this semiarid portion of the plains, "drought" was a relative term. Compared with the eastern plains, the western lands lacked the precipitation to maintain Corn Belt farming practices. Since the pioneers first came to the western plains, farmers and public officials attempted to adjust both their agriculture and expectations to the often inadequate rains of the area.

Three generations later, farmers still hadn't fully adjusted to the environment. The region's climate deceived and often impoverished farmers. In cultural terms, they faced a want of both rains and opportunity. In scientific terms, much of the plains lacked dependable rainfall. The National Weather Service describes drought "as a period of 30 days or more during which rainfall was deficient, no day in the period receiving more than one-quarter inch of precipitation."[21] Eventually during the 1930s drought affected all areas of the Great Plains. For the region as a whole, between 1930 and 1936 the drought pummeled western and central Kansas, northeastern Nebraska, nearly all of South Dakota, and parts of southern and western North Dakota.[22] It was worse in the western third of Kansas, for example, which suffered from drought conditions from the fall and winter of 1932–33 until rains relieved parched soils in October 1940.[23]

The timing and locale of drought conditions during the 1930s defy generalization. In a given month, fields in western Nebraska received adequate moisture, while the soil in the eastern Dakotas remained parched. In addition, the needed moisture that determined whether a particular crop was successful might come at just the right time during a growing cycle—or not come at all. Other factors came into play, such at the hot winds that dried up topsoil moisture, poor soil conditions that allowed precious rains to run off, or farmers lacking the experience, knowledge, capital, or machinery to make the best use of scanty moisture.

The one fact that *can* be stated is that the plains drought was recurring, and to the farmer it came with impeccable timing to ruin the crops. When rains came, they always seemed to fall after crops were beyond salvaging, and so were of little use, since fields invariably dried up before the next year's harvest. A report from the Atchison, Topeka, and Santa Fe Railway in July 1934 held true for any time or area of the plains during the 1930s. "An amazing thing about the drought this year is the fact that it has been broken frequently," the report stated, "but refuses to stay that way." It continued, "Light rains bring to the parched soil moisture sufficient to add new life to crops and make them start growing again; but immediately subside, and more drouth [sic] comes to pester the lives of the farmers, and scare town folks."[24]

Such dry spells lasted throughout the 1930s. For example, in 1931 a Nebraska Farmers Union official reported that the sun burned his crop so badly, "I could put the whole of the crop in my old straw hat (which is left over from last year) and not stretch the band."[25] Six years later the chairman of the Butte County, Nebraska, County Board simply reported to the governor that "the wheat crop is gone." He recounted how one farmer in the western Nebraska county harvested twenty acres of parched wheat, only to yield a pathetic total of nine bushels.[26] Under normal conditions, the entire wheat field should have yielded at least 260 bushels.

Surely the most evocative aspect of the drought of the 1930s in the American Great Plains was the appearance of the Dust Bowl. Even under normal conditions, precipitation in the semiarid western plains amounted to only enough to raise dryland crops such as wheat. Blistering heat and scant rainfall during the thirties proved that farmers had not mastered nature. During the early years of the decade, drought combined with questionable farming practices transformed a large region in southwestern Kansas, northeastern New Mexico, and the panhandles of Texas and Oklahoma into what appeared to be a desert. Impoverished farmers were defenseless, since they could not afford to pay fuel, labor, and machinery expenses to practice soil conservation through strip-cropping, terracing, and listing their fields.

Stories abounded of hot winds that rivaled the tall tales of the Old West whipping dust into the atmosphere. Newspapers reported that a Phillipsburg, Kansas, janitor swept five hundred pounds of dust from his high school after a storm.[27] Along with Kansas, drought scorched and dust storms scoured parts of Nebraska and South Dakota. An agronomist for the University of Nebraska College of Agriculture in 1936 predicted dust storms "the like of which Nebraska has never witnessed," should the rains fail to come in 1937.[28] In South Dakota, witnesses to the storms cited their cruel impact on both animals and humans. T. H. Ruth of the state agricultural department noted that horses and cattle perished not only from a lack of water but also from sand and dirt balls in their digestive tracts.[29] Years later, a South Dakota farm woman recalled that during storms dust seeped into her bedroom, "and the dust would be so thick on the bedspread you couldn't tell what color it was." Her daughter remembered filling a bucket of water from the outdoor cistern. By the time she reached the farm home, it was full of dirt, making the water undrinkable.[30]

Insect pests were the second affliction that troubled plains farmers during the 1930s. Although other pests such as chinch bugs caused extensive crop failure, grasshoppers were the most harmful. Travelers journeying across Kansas in 1936 noted how plagues of grasshoppers had devoured corn plants back to their stalks. Farmers in southeastern and south-central Nebraska cast anxious eyes skyward that year as grasshoppers flew over the region, their wings looking like "yellow cellophane" from the ground. Grasshoppers inflicted serious damage across the state.³¹

Torrents of grasshoppers also overran swaths of South Dakota. Relief investigator Lorena Hickok reported in November 1933 that after the insects came to South Dakota, entire fields of grain looked as though they had been freshly plowed. Hoppers cleaned the bark off trees, and residents learned by experience not to leave their wash on the clothesline lest grasshoppers eat it off.³² The flood of grasshoppers left vivid memories for South Dakota farmers. Ella Boschma, who farmed with her husband, Leonard, near Springfield, South Dakota, recalled applying banana-scented grasshopper poison by hand between the crop rows. However, it was to no avail. "You'd kill some, but it seemed like there were more that came. [People] said [other grasshoppers] came to the funerals." Ben Huggins, a farmer living near Geddes, South Dakota, saw the thorough damage done by the insects. They cut a stalk of corn like a scythe, then sucked the moisture from the corn roots, leaving the plant dead in the ground.³³ The damage from grasshopper invasions did more than ruin a year's crop. By leaving the land without cover, their visits often led to soil erosion.

With all the problems that farm families faced, it would be easy to forget the attractions of farm life for these people. But farming offered a sense of community that no other line of work provided. Rural neighborhoods were connected by links of exchange, interrelationships, and mutual assistance that lessened the harsh economic and environmental problems of the 1930s. For example, although combined harvesting machines were taking over plains harvests during the thirties, many farmers still joined together in threshing crews. These crews consisted of around ten farmers who hauled bundles of wheat from the field, threshed the wheat, and scooped the wheat into grain bins. Neighbors worked together on these crews until they threshed the entire crop. Harold Clingerman, a Nebraska farm manager, recalled how farmers settled the differences in labor with each other. "We got your wheat down in half a day," one would say to his neighbor, "but you worked all day in my wheat and barley, so I owe you a half day. Call me when you need help." Women on the farm fed the threshing crew, and it was considered rude to turn down a place at the dinner table.³⁴

There were other, less tangible benefits of farm life that didn't show up in USDA reports or census data. Farmers treasured being their own boss and not having to follow orders from someone else. Farm men and women took pride in doing a good job and seeing their handiwork pay off for them. Whether it was piling a haystack to deflect rainfall, knowing how to get the most out of a flock of chickens, or keeping a presentable home and farmyard, they got satisfaction from using skills they had learned from their parents and neighbors. Farmers who stepped outside on a bright summer morning to find their own farm before them, ready to be worked, didn't have to ask why other farm families remained on the land as long as they could. Individual farm families struggled to keep a farm that was their home, their work, and their way of life.

Business Strategies for the Landed Poor

The treacherous environmental conditions and low commodity prices of the thirties combined to make insolvent even the middle-income men and women with substantial property. For instance, the ample amounts of property and livestock owned by the middle-income farmers visited by Lorena Hickok had not kept them from falling into insolvency and poverty. It was probably small comfort to these farmers that nearly all agricultural operations, large and small, suffered from a fall in income beginning in the early 1930s. Drought, grasshoppers, and high debt rates hit all sectors of the countryside. The American farmer had produced huge stockpiles of American corn and wheat and vast numbers of hogs and cattle which still flooded the markets. Despite New Deal agricultural programs designed to cut the surplus of each commodity, surpluses crept upward to predepression levels by the later 1930s.

Despite these problems, some plains farm families achieved, if not prosperity, more economic security than others. These were most often large farmers with large incomes who serviced their debt. The more successful farms attained greater economic stability by becoming more efficient, by controlling farm costs, achieving greater productivity, and by earning non-farm income. The individual strategies these plains farmers used to persevere during the 1930s deserve attention.

Farmers used various means to get more from the land. Agricultural productivity arises from the farmer's managerial ability and the appropriate size and use of the land and its resources. The farm business was a combination of luck, hard work, and shrewd planning. Everything from the selection of crops and livestock to knowing how to deliver them to market terminals meant the difference between profit and failure.

An obvious strategy was to farm big. In 1932 the South Dakota Agricultural Experiment Station examined profitability in the state's spring wheat belt based on actual farm records. The study constructed three models of farm sizes ranging from large (800 acres), medium (480 acres), and small (240 acres) operations composed of owned and rented land. Not surprisingly, the report found larger farms to be the most profitable because they most efficiently used capital, labor, land, and their productive potential. The goal of farming profitably was to spread expenses across the most number of acres possible. Tractor farming appeared to be most sensible on a larger farm able to distribute farm power expenses across more land. In addition, the study cited that larger farms had a lower proportion of "wasted" land. Most farms, large and small, have to remove land from production for use in roads, buildings, and a dwelling. Larger farms were able to spread this "waste" across more acreage.[35]

Success Stories

Some farmers achieved limited economic security during the Great Depression. How did they make it? The South Dakota Agricultural Extension Service suggested making judicious use of capital, investments, and resources on large amounts of acreage to make the most profits.[36] Richard Bremer, in his history of central Nebraska farms, proposed that certain types of farmers survived and succeeded. Those who were able to keep their operations through drought and depression were mortgage-free. They were old enough to have paid off their mortgages or young enough to have secured land without acquiring debt. Bremer describes many of these as "conservative" farmers with a strong "security orientation." During the economic doldrums of the 1930s, these farmers may have placed themselves outside the race to increase their productivity through larger-scale agriculture. By avoiding mortgages and the capital investment in machinery and outside labor, they showed their own adaptation to the Great Depression. In contrast, Bremer finds "progressive farmers" had to scramble by cutting operational expenses to meet their mortgage payments.[37] However, conservative farmers may have achieved their security at the expense of higher income and long-term adaptability to the changing farm economy.

Those who lost their homesteads, according to Bremer, were pummeled by the Great Depression while still paying off debts or acquiring land and mortgages. Bremer also found that fertile land, good weather, managerial skills, family loans, and government agricultural programs also were helpful.[38] Likewise, historian Mary Hargreaves found that plains farmers with low acreage, few cattle, little land devoted to feed crops, and a great proportion of their land in pasture often failed.[39]

These are generalizations. Two accounts of individual farm families offer personal portraits of struggling farmers. Mern and Ada Wall, of Grand Forks County, North Dakota, had a successful depression-era farm. Mern and his brother Earl pooled resources to run an eight-hundred-acre farm. They had a $5,000 mortgage on 320 acres and they leased the remainder of the land. Through shrewd management and keeping the debt service at a minimum, the Walls saved cash for maintenance and investment in the farm. When the rains returned, Ada and Mern had kept their farm and were in a good position for prosperity.[40]

Deborah Fink in her study of Nebraska farm women cites another example of a farm couple who managed their way through hard times. "Sam" and "Nora" of Boone County, Nebraska, were married in 1931. During the 1930s Sam managed a trucking business, service station, and farm implement company, while Nora ran the farm and kept the books. Through hard work, family support, and juggling time and resources, they were able to remain in Boone County at a time when many farm couples hit the road in search of work.[41]

Diversification and the Farm Crisis

Farmers and farm experts were aware of the problems of western agriculture. Drought and pestilence leading to crop failure, debt, and low income were the immediate crises in the plains during the thirties. These problems affected all who made their living off the land. The long-term agricultural and economic trends of the times also placed plains farmers at a special disadvantage. While the solution for a long-term paying operation was to transform into a mechanized, highly productive, and highly capitalized enterprise, many plains farmers lacked the acreage and farm infrastructure to make the jump into "modern" farming.

In response to both the hard times of the 1930s and the expensive, mechanized, and newly productive agriculture, county farm agents promoted diversified farming. Many believed that dependence on one-crop farming, especially in gambling that the price of wheat would return to the prices of the World War I era, was foolhardy. County agents encouraged farmers to broaden their production and raise more of their own food and fuel for home consumption. Through this, they said, farm families could count on a variety of products, making them less susceptible to the unpredictability of King Wheat and King Corn commodity prices. However, diversification was difficult for farmers in semiarid areas, since most farms were established for raising small grains and a few livestock profitably. They believed that the wealthy family down the road got that way through corn, wheat, and cattle. Also, most farmers in the semiarid western plains believed that the area was most suitable for wheat and cattle. Thus, one agricultural economist said, "I haven't seen a real honest to gosh paying diversified farm west of the Red River and surely not west of the Missouri."[42]

The idea that diversified farming was the solution to farm problems was flawed. First, the Great Plains as a region already grew several crops. In 1940 only 40 percent of all cropland in the region was in either corn or wheat.[43] Furthermore, farmers resisted efforts to diversify their produce for economic and personal reasons. It was expensive to convert from an operation set up to grow only a few types of crops. For example, county agents urged farmers to build silos to store their own livestock feed to cut costs. Yet these were comparatively expensive and time-consuming to build and replenish.

Wheat growers in particular refused to diversify. For example, though prodded to also raise cattle for dairying, they found it unprofitable to milk smaller dairy herds and difficult to find buyers for their cream in the 1930s. Historian David Danbom writes of the many reasons North Dakota spring wheat growers refused to expand dairying beyond subsistence needs. Despite state encouragement, North Dakota farmers knew that they were too far away to supply urban milk markets, unlike Wisconsin and New York state. That left them with the less-lucrative butter and cheese market. Furthermore, they were spring wheat farmers, plain and simple. Increased dairying took away their leisure time and required expertise they lacked. In Danbom's words, skeptical farmers saw "it was foolish to turn some of the world's finest spring wheat land into pasture [for dairy cows] and vastly increase the labor requirements of the farm just to engage in the least profitable aspects of the dairy industry." To large-scale spring wheat growers, increased dairying meant poverty-induced self-sufficiency.[44]

Likewise, in other areas of the semiarid plains, winter wheat was still king, and prospects for diversification into other crops were few. Earl Bell, in his study of Haskell County in the southwestern corner of Kansas, found that in the early 1930s farmers planted wheat in 94 percent of the cropland. Wheat growers here saw diversified farming as merely a form of subsistence agriculture joined to their cash-crop operation and was therefore not worth their time.[45] Furthermore, running a successful diversified operation took extensive management skills, which many farmers lacked. Finally, the early New Deal agricultural program under the Agricultural Adjustment Administration (AAA) rewarded those who raised a limited variety of produce. In the plains the first AAA compensated only those farmers raising corn, wheat, and hogs for cutting back on their production.

Diversification strategies addressed important agricultural, economic, and environmental problems in the plains of the 1930s. For a region awash in debt and surplus grain, diversified agriculture offered plains farmers alternatives to monoculture. But rejecting large-scale commercial grain farming would have required a substantial agricultural, economic, cultural, and ideological shift. Such a change meant abandoning opportunity, higher income, and increased productivity on mechanized farms. It meant the deferment of cars, labor-saving appliances, machinery, and consumer goods that plains farm families grew to expect. Therefore, despite calls for a planned conversion, the primary goal for farmers wasn't diversified farming and raising food for the dinner table. The real issue was maintaining and expanding the *scale* of productivity. However, this was an unreachable goal in the face of the environmental and economic hazards of the thirties.

Rural Poverty

One of the impediments to transforming farm operations was poverty. But determining poverty involves a value judgment. It is difficult to find an official definition of "rural poverty level" in the 1930s by contemporaries or historians. In 1934 the Brookings Institution found that 54 percent of the nation's farm families (comprising 17 million people) made less than

$1,000 per year, which the institution took as a sign of America's poverty. In addition, the South Dakota Agricultural Experiment Station indicated that one of the causes of "depleted resources" was an "insufficient farm income" of less than $1,000 annually.[46] This definition included not only cash income but the value of goods and services created or bartered from the farm.

By this definition, in 1940 the Great Plains contained 213,000 low-income farms, comprising nearly half of all the region's total. With an average of 4.5 persons in each farm household, 960,000 people lived in poverty in the rural plains before World War II.[47] But rural poverty was well established in the plains and United States before the Great Depression. In 1930, before the countryside felt the full weight of the economic slump, 88,000 farms, or 17 percent of all plains farm families, made less than $1,000 annually. Yet the plains was well off compared to the United States as a whole.[48] Inflated by the extraordinary, intractable poverty of the South, 47 percent, or nearly 3 million American farm families, lived in poverty in 1930. By 1940 that number grew to 4 million farms, comprising two-thirds of the nation's total. Yet lower rural poverty rates in the plains did not necessarily mean higher incomes. While average gross income of plains farms was $1,410 that year, that of American farms was $1,840. Therefore, although the plains had a lower proportion of farmers living in indigence in 1940, it also lagged behind the general farm income in America as the nation began to emerge from the Great Depression.[49]

In contrast to rural standards, the Brookings Institution set the "poverty line" at $2,000 for nonfarm American families. Significantly, the average annual earnings for those employed in sales ($1,380), public education ($1,440), and federal workers ($1,900) rested below the poverty level in 1940. Careers in sales, education, and government were traditional routes for many into a middle-income standard of living. Of course, the more important problem for nonfarm workers during the Great Depression was rampant unemployment, which reached 48 percent in 1933. That year, nearly 13 million unemployed Americans swelled the relief rolls. Between 1931 and 1940 the jobless rate among nonfarm workers never fell below 21 percent.[50] This intransigent problem of unemployment in American towns and cities committed New Deal planners to keep as many farmers in the countryside—and out of the city soup lines—as possible.

Borderline Farmers and Their Farms

Yet my focus is on neither the urban unemployed nor destitution in the plains countryside, but rather the farm families on the margins of poverty. Lorena Hickok visited such struggling though propertied families in late 1933. They were the true targets of the rural rehabilitation program. While a huge proportion of plains farm families lived in severe poverty, with little opportunity to remove themselves, many struggled to attain middle-income status. These farmers lived on the borderline between indigence and the limited security of middle-income farmers. These were "borderline farmers," grossing between $500 and $1,000 annually during the 1930s.

Farmers of all income levels grappled for economic security during the Great Depression. Caught in surroundings hostile to raising crops or livestock, borderline farmers worked hard to survive with their limited resources. Certain parts of the Great Plains had greater proportions of borderline farmers than others. In 1930 these areas were in northwestern North Dakota and in north-central South Dakota. In addition, in a band of Kansas counties reaching south from Kansas City to the Oklahoma state line, and counties in the north-central part of the state, farmers averaged between $600 and $1,000 annually.[51] The economic and environmental forces of the 1930s doubled the percentage of borderline farmers among all farmers from 10 percent to one-fifth of all plains farmers by 1940.[52]

Data from the 1930s shows how these borderline farm families differed from their wealthier neighbors. The most important determinants of a farmer's income potential were acreage and value of the operation. What was the average size of a borderline plains farm? Unfortunately, it is difficult to find data giving the number, use, and value of acres by income level. Even if such data existed, not all land, even within the same county, was equally productive. For example, a two-hundred-acre farm in Douglas County, Kansas, in the Corn Belt, was more valuable than one in Stark County in semiarid western North Dakota, with the same acreage. Add to this the variables of soil quality, commodity market fluctuations, government agricultural subsidies, investment in fertilizers, machines, labor, and livestock, as well as the managerial ability of the farm operator, and you have an income unpredictable solely on the basis of acreage. Delineating farmers by their farm's value is also unhelpful. A farm owner, for example, was often better off than a tenant, even if their farms were equally valued.

Despite the shortcomings of grouping farms by acreage and total value, this data places these borderline farmers within the context of the plains countryside. In a multicounty 1935–36 study conducted by the National Resource Planning Board (NRPB) based in southwestern Kansas and eastern North Dakota, borderline farmers ran operations three-quarters the size of those of general farmers.[53] Another rough indicator of the farm's income potential was the value of the individual homestead and surrounding land. The value of farmland and buildings for borderline farmers in the NRPB study was 17 percent less than the value of all farms.[54] These figures suggest that with substantially smaller and lower-valued operations, borderline farmers stood apart from their neighbors. By appearance and fundamental criteria, borderline farmers comprised a distinct, troubled sector of the countryside.

Plains borderline farmers were especially vulnerable to both the long-term agricultural trends and the brutal economic and environmental forces of the 1930s. These were peripheral farmers within a peripheral region. In fact, the region attracted aspiring men and women because families could afford to start a successful farm there. In the eastern, more humid plains, borderline farmers hoped to prosper through diversified, smaller-scale agriculture. They were also drawn to the western, semiarid lands by large-scale, mechanized farming. However, this was an inauspicious time for both subregions. While the agricultural sector became less significant within the national economy, these aspiring peripheral farmers became less economically viable in the plains farm economy.

The 1935–36 NRPB survey of North Dakota and Kansas farmers suggests how expenses differed by income level. Generally, middle-income farmers, grossing between $1,000 and $2,499 annually, had higher expenses. These outlays boosted the operating costs per acre. However, middle-income farmers were able to more than offset increased costs per acre with increased income. Borderline farmers were not.

Another way farmers increased their economic security and lowered their debt was to cut back on operating expenses as much as possible. Obviously, all farms have operating costs. In general, the borderline farmers surveyed by the NRPB had the same types of operating expenses as middle-income ones. The problem for borderline farmers was that many essential expenses, such as livestock feed, crop seed, rent, and loan interest and refinancing charges, were the same or higher for themselves compared to larger operators. Unfortunately, borderline farmers had assumed many of the expenses of more prosperous farmers but without the income to pay for them. For example, the costs of live-in household and agricultural labor were generally the same across income levels. Overall, most borderline farms spent less on expenses more easily foregone, such as livestock, machinery, and tool purchases.[55]

Correspondingly, within America's Farm Belt, most borderline farmers were unable to lower their expenses by mechanizing. In the west–north-central region, which includes the plains states plus Minnesota, Iowa, and Missouri, only one-third of the borderline farmers used tractors; this was three-quarters the rate of the general farmer in the region.[56] Therefore, borderline farmers reduced their initial operating costs by spending less on machinery and tools. However, by cutting back on critical investments, they may have missed their opportunity to cut costs—and their chance to build a viable, profitable operation.

After cutting expenses, farmers supported their business through non-farm income. In 1939 one-quarter of all plains farmers labored away from their land for an average of seventy-nine days that year. Seven percent worked one hundred or more days off their home place. Where did they work? Forty-four percent reported working on other farms, while 63 percent reported working in nonfarm positions in 1939.[57] In this climate income and gender determined nonfarm wages. NRPB–surveyed farm husbands of all income levels worked off their land at the same rate. However, once they got off their land, borderline farmers received substantially less than the average farmer. Borderline farmers were younger than general farmers and probably lacked the skills and contacts that their more substantial neighbors had. Furthermore, since plains borderline farmers had less machinery than the average farmer, they missed the opportunity to contract themselves and their machinery out to others. Finally, if a borderline farm man lacked the money to keep his car running, he lost the chance for short-term employment in town. The NRPB survey indicates few farm women of any income level worked away from their own kitchen and farmyard. Yet the survey failed to note that some jobs that farm husbands took on other farms came with the understanding that his wife would perform domestic jobs for his employer.[58]

Finally, and most important, levels of farm productivity distinguished the prosperous farmer from his borderline farmer neighbor. NRPB–surveyed middle-income farmers in Kansas and North Dakota made 26 percent more income per acre compared to borderline farmers.[59] Federal census figures showed that while plains borderline farmers comprised 20 percent of the agricultural population in 1939, they produced only 10 percent of the agricultural products. On the other hand, farmers grossing $10,000 or more annually represented only 0.7 percent of all plains farms, yet they produced 12 percent of the total value of farm goods.[60] Therefore, based on productivity, the wealthiest plains farmers could have replaced the borderline farmers. Following the logic of the accounting ledger, the plains borderline farmer was expendable, and during World War II, many larger farms absorbed these smaller operations.

The plains borderline farmers had a specific place within the modern agricultural market economy. These were small-scale and commercially oriented men and women who remained strongly committed to farm life. The men devoted much of their labor to the farm, while the women spent all their working hours on the homestead. Because of the agricultural trends of the 1930s, both husband and wife were less productive than their counterparts on larger farms. During the Great Depression their diminished income placed them on the margins of poverty. Most borderline farmers were not practicing subsistence agriculture, since they paid for outside labor, feed, machinery, and fuel. Most significant, they depended on outside credit for their operations. Borderline farmers in the NRPB study owed an average $92 in interest and refinancing charges alone in the mid-1930s.[61] They were trying to not only maintain commercial farms but increase their scale of operations as well.

The data offers the cold facts about borderline farmers' business orientation. But what about their personal feelings about their way of life? It is difficult to determine their exact goals for their operation. However, O. J. Maurer's thoughts in his letter to Franklin D. Roosevelt probably fit those of borderline farmers. In December 1932 the 45-year-old Maurer of Pukwana, South Dakota, wrote president-elect Roosevelt to explain his troubles. Maurer had a large farm of 640 acres and a mortgage of $12,000. He was $3,000 behind in his mortgage payment and taxes. "Under the present conditions with prices so low I am certain that I cannot carry on," he wrote. "The interest and taxes are more than I can possibly raise in one year. . . . if I am to give up our home it would mean the loss of practically $60,000 or the savings from toil, of two generations as this farm was left to me by my father."[62]

Like many businessmen during the early 1930s, Maurer was heavily in debt and in danger of losing his operation. But most important to him was his life as a farmer. If creditors foreclosed on Maurer, it meant more than the loss of his business; it meant the loss of his home and a way of life. Such a loss would wipe out not only his years of labor but those of his father who left the land to him. Like Maurer, borderline farmers were both entrepreneurs and people of the soil. Tens of thousands of insolvent plains farm families struggled to keep their enterprise afloat to preserve their businesses and their homes. They were businessmen and women who used commercial agriculture to maintain and improve their way of life. They worked to support their families in healthy and prosperous surroundings and to acquire and preserve a viable farm to pass on to their children. These were conservative goals. However, borderline farmers had to live and operate within a high-risk, modern commercial economy to achieve these goals. Although the market economy frustrated them, they did not try to change the structure of modern large-scale farming. They wanted to succeed within it rather than replace it.

Borderline farmers were ill-suited to meet the demands of plains farming during the drought and depression of the 1930s. More substantial farmers achieved greater economic security by managing larger farms at lower costs, by earning nonfarm income, and by increasing or at least maintaining production. Most borderline farmers were left out. These small business men and women had managed to increase their productive costs, while remaining too unproductive or too unprofitable to pay for their expenses.

These borderline farmers were not necessarily unfit. Rather, they were out of their league. Many smaller farmers and those free of debt rode out the depression if they had adequate rainfall. Farmers who avoided outside credit achieved some economic security. However, this came at the expense of a diminished standard of living. Modern agriculture during the 1930s required expanded acreage and mechanization, and the commercial credit to pay for them. Borderline farms were large enough to aspire to such ends but not profitable enough to pay for the means. These Plains farmers failed to keep up with the growing scale of agriculture and were becoming sunset farmers. There was no longer room for the smaller farmer in the American Plains. Yet these borderline farmers still demanded aid from the federal government. They were answered with the rural rehabilitation program.

2

Farm Tenancy in the Great Plains

Tenancy and Opportunity in the Pre-Depression Plains

The rich brown and black soils of the Great Plains had made a siren call to aspiring landowners since the pioneer days. Despite the proven dangers of insolvency, erratic income, and environmental hazards, tens of thousands of farm families came to the region to enter into the venerated class of property owners. However, by the 1930s a near majority of farm families lacked title to the fields and farmyards in which they toiled. In fact, the great majority of plains borderline farmers, who made between $500 and $1,000 annually, shared a common trait: they did not own the land they cultivated. The New Deal's rural rehabilitation program, run by first the Resettlement Administration then the Farm Security Administration, assisted borderline farmers who were mostly tenants. In search of some type of economic security and expansion, borderline farmers chose to rent or had to rent their land because of a shortage of capital or adequate affordable land.

Many regretted this, for, like millions of Americans before them, they saw landownership as their ultimate goal. The 1930s may have been the last time that many Americans aspired to be farm holders and equated landownership with status and opportunity. One woman reflected this sentiment in a breathless letter to the *Wichita Independent* in 1937. She wrote that with land "I would be the happiest woman, wife and mother, if God would only grant. . . . I would just about go raving crazy with joy if a chance could come like that, and an opportunity to live until we could raise something." She continued, "I'd feel that I could raise my head up and say 'Now look at us.' If they would give people a chance with small farms, cows, horses, chickens, and pigs." The woman ended, "Please, won't you help us go on one. I'd certainly be the happiest human."[1]

Ideally in the United States, the citizen owns his or her property, free of debt or obligation to another. The propertied American takes pride in ownership, particularly of land. The owner cares for it, invests in it, builds up its productivity, and profits from it. From the land, the possessor receives not only a livelihood but status and identity. In the early American republic, property was often the prerequisite for full and free citizenship.

During the 1930s the majority of plains residents believed the farm owner made not only a better farmer but a better citizen. Correspondingly, many feared a countryside filled with renters. In 1937 the South Dakota State Planning Board echoed many farm observers when it asserted that "Tenancy is usually considered undesirable because under the system prevalent in the United States it frequently leads to the exploitation of the soil, the deterioration of farm property, and a higher mobility rate in the farming population." The results, claimed the board, were "instability in local community institutions," and tenancy meant "a greater degree of economic insecurity and a relatively low economic status."[2]

Like the nation's founders, many still believed that a nation of small landed proprietors formed the bulwark of a healthy society and democracy. Yet even George Washington doubted the benefits of landownership for all. A landowner himself, Washington sought tenants for his properties. He commented that many immigrants made "precipitate purchases" of land before they were ready for the responsibility. Washington advised that "if on advantageous terms they could have been first seated as tenants, they would have had time and opportunities to become holders of land, and for making advantageous purchases." But Washington warned that "it is so natural for man to wish to be the absolute lord and master of what he holds in occupancy, that his true interest is often made to yield to a false ambition."[3]

Despite these misgivings, however, a fundamental belief in the benefits of landownership persisted in America. This explains why the rise in the number of farm tenants was so troublesome between the two world wars. On the eve of World War II, 2.4 million American farmers, including 200,000 plains farmers, rented their land. That meant tenants rented 39 percent of American farms, and half of plains farms.[4] Despite some evidence of a stable class of renters, the high tenancy rates and large numbers of renters conjured images of low income, diminished living conditions, social instability, and soil erosion in rural America. Therefore, correcting the high rates of tenancy became a morally compelling goal during the 1930s.

Farm tenancy was a very complex issue. Renters ranged from some of the wealthiest farmers in the plains to many of the poorest. Renting itself was quite diverse. Landowners had different reasons for leasing their acreage, different relationships with tenants, and they expected different types of leases and payments. The tenants themselves had their own reasons for renting. Some used farm tenancy for their own needs, others were used by it. Although many plains farmers cultivated both owned and rented land, I am comparing only full owners and tenants without their own land. Neither form of land tenure—ownership nor tenancy—automatically led to economic security in the plains. There were cases of both successful renters and many, many struggling farm owners. However, borderline farmers and tenant farmers shared similar key traits. Both groups were rural entrepreneurs who often lacked the capital, land, income, and general wherewithal to claim any kind of social and economic safety during the Great Depression. Therefore, many tenant borderline farmers were less secure than owners, for they lacked even the shelter of landownership.

Many critics during the 1930s viewed tenancy as a sign of social and economic injustice. They associated low economic and social status with farm tenancy, although many renters were quite successful. Many tenants were related to the landowners and expected to inherit the property. Furthermore, for ambitious and better-off tenants, renting land was often an economic response to economic and agricultural problems. Since tenants' limited capital was not tied up in mortgage debt, they could invest in increasing the scale of the operations. Farm tenancy was a potential business strategy for young farmers with the resources and the skills to take advantage of times of high operating costs and high commodity prices. Unfortunately, for a great many farmers, tenancy was not a profitable strategy but was the least expensive way to enter or remain in farming. Unable to meet high mortgage expenses, a great number of plains tenants must have rented just to stay in the game.

Depression-era agriculture meant thousands of young tenant farmers were frozen out of farm ownership. In response, even moderates proposed radical solutions to this problem. They questioned the basis of practices followed in the plains since the pioneer days—the right of farmers to buy, steal, rent, or squat on as much land as possible for speculative purposes, for inheritance, or to increase the scale of their operations. Conservatives replied that farm tenancy was a mechanism of market forces that winnowed the winners and losers in plains agriculture.

The growth in farm tenancy rates was a long-term trend in the United States and the plains states. From 1880, when the federal census first published land tenure data, until 1930, both national and plains tenancy rates rose. They leveled together at around 40 percent during the 1920s. However, after 1930, while national tenancy levels dropped, those in the plains shot up to nearly 50 percent in 1940. These high farm tenancy rates made the plains stand out within the nation. From the Canadian border to the Oklahoma state line, the region's counties had at least half their farmland under lease. The exception was the Sandhills region of Nebraska where ranching dominated.[5]

Therefore, from 1880 onward, tenancy rates in the plains expanded steadily despite the huge supply of unclaimed land. There were several explanations for this. In the late 1800s farm tenancy increased because of the disappearance of cheap land and the mounting capital costs of farming. In addition, many of the original settlers were retiring and renting their land to their children. This trend continued into the twentieth century. The percentage of renters among plains farmers exploded after the World War I when skyrocketing land prices and general inflation blocked the route into farm ownership.

As the farm tenancy rate increased to 40 percent of all American and plains farms during the 1920s, farm observers grew quite concerned about tenancy's effect on the countryside. Kansas farm tenancy was a case in point. One survey of Sedgwick County in south-central Kansas called tenancy a negative influence on rural community life. The Interchurch World Movement cited that over half of Sedgwick County farmers rented their land. However, only 12 percent of the farm families who attended church were tenants. The survey charged that while many renter parents sent their children to Sunday school, they skipped religious services themselves.[6]

Farm journals such as *Country Gentleman* took a great interest in the allegedly negative impact of farm tenancy in Kansas directly after World War I. In January 1919 reporter Harry O'Brien visited Woodson County, Kansas, east of Sedgwick County. Investors from Kansas City or Wichita owned most of the rented farms, he reported, and they rarely visited their properties. O'Brien concluded that Woodson County, with 42 percent of its farms operated by tenants, was "a fine example of what a farming community ought not to be." He noted that most rented farms were for sale and leases lasted only one year. Most owners hoped to sell out at the year's end. "Knowing this," O'Brien wrote, "the tenant tried to get all he can out of the farm in one year. He expects to move on, so why worry how he takes care of the place?" O'Brien saw a countryside dotted with poorly run, soil-depleted farms. Tenants refused to look after the county roads, and their homes were in a "deplorable" state. Conditions such as these were typical for eastern Kansas, according to O'Brien.[7]

Some blamed farm tenancy on large outside landowners who snapped up rented farmland. At the end of 1919, *Country Gentleman* ran an article by Governor Henry J. Allen of Kansas condemning absentee landlords. These "land hogs," according to the governor, cheated tenants out of the opportunity to purchase their own land. This injured the rural community and endangered the stability of the nation as well. Tenants invested little in their countryside when at the end of the year's lease they moved again. The tenant "has no welfare interest in the community," wrote Governor Allen. Furthermore, the renter was "an easy prey to the class-hatred propaganda of the i.w.w., the Bolsheviki and the other class-minded organizations that seek profit out of his discontent."[8]

Such observations of 1920s Kansas renters, their supposed lack of community participation, high geographic mobility, and the shabby appearance of their farms painted a negative image of farm tenancy in general. To a nation fearful of change, tenancy embodied the instability of the era. This was a time when thousands of families were cutting their ties with their rural homes and moving to the towns and cities. While the city gleamed with excitement and opportunity, many people in rural America worried that the countryside was falling behind the rest of the country in income and opportunity. This fear reflected an ambivalence toward farming. Agricultural publications and county agents urged farmers to run their operations in a "businesslike" manner. However, the economically logical approach to post–World War I farming was to expand operations while cutting liabilities. Renting land was one way of cutting liabilities, but it also entailed cutting ties to the local rural community. It was a dilemma that troubled many plains observers.

However, some Kansans during the early 1920s defended farm tenancy as a means for a farmhand to become a farm operator. In 1920 a Brown

County renter appraised the situation. In recent years cash rents had increased, as had the share of the crops the tenant surrendered as part of the lease. Also, landlords were furnishing less and less to the tenant to run the farms. Still, for young farmers with little capital, renting was the best means for acquiring their own operation, boosting income, and transforming themselves into owners. "At its best and at its worst," he claimed, "tenancy presents no more faults than exist in American farming. . . . Let us just remember that tenancy is the farm hand's industrial opportunity in the country." The article included a cartoon in which a farm worker was accompanied on his left hand by a brazen flapper representing "The lure of the City," and a sober elderly man on his right representing "Tenancy." The flapper pointed to the flashy "Big Wages and the Gay Life" of the city. The elderly man pointed to "The Chance to Come into the Land-Owning Class" at the end of a rainbow.[9]

Others pointed to ideal examples of land tenancy in Kansas. At the beginning of the Roaring Twenties, *Breeder's Gazette* noted a model owner-renter relationship that lasted despite the boom-and-bust cycles that rocked the plains farm economy. Frank Prather of Lyon County, Kansas, had the same tenant family for twenty-seven years. Prather and his tenant were partners on the land. They owned the livestock and machinery jointly, and after expenses the two men split the profits equally.[10] The more positive views of renting provided a reassuring picture that was more amenable to the plains and the market economy. By presenting farm tenancy as an opportunity for ownership or as a landlord-renter limited partnership, apologists of tenancy justified its growth during the 1920s.

"Land Hogs" and the Scale of Agriculture

The crash of 1929 hit borderline farmers who rented their land at an especially inopportune time. If they aspired to ownership, they faced a decade of cruel disappointment. The negative trends endemic in farm tenancy during the 1920s and 1930s clobbered these aspiring men and women before they were able to convert from renter to owner. Farmers and farm observers of the era debated the cause of increasing farm tenancy during the 1920s and 1930s. Many blamed higher tenant rates on the large landowner, who snatched up land before renters could afford to purchase it. From west-central Kansas cries arose against these "land hogs" that Governor Allen castigated after World War I. Bert McConachie of Jetmore complained that young men couldn't get a start in farming because of the "wheat hogs" who acquired the land to become landlords.[11] A Catholic priest in Russell County, Kansas, advised President Franklin Roosevelt that "there is an evil arising and increasing in this community as well as in other farming communities of the West. Too many farmers are farming too much." He charged that many landlords were tossing their tenants off the land and farming it for themselves. The reverend submitted the radical solution of regulating farm size.[12] One farm owner in Pretty Prairie, Kansas, offered his own views on the problem. America, wrote L. R. French, was becoming a "landlord owned nation." French continued, "I see no logical reason why I should own the farm of my neighbor and charge him for its use. The Creator of the world made the fertile land and gave it to the people of the world to use. . . . I believe it unfair for one individual to charge another for the use of land in excess of what one man can reasonably farm. I should have the opportunity to own and operate all the land that I can reasonably use, just as long as I do not infringe on the right of my neighbor to do the same." French argued that "landlordism" had driven up land prices and suggested a graduated tax on operations larger than the "average family type farm" to solve the problem.[13]

One young man wrote that he was effectively barred from farm tenancy. In 1938 Frank Bean of St. John, Kansas, had the unenviable job of driving a truck for a rendering plant. Like many men, Bean was looking for opportunity when it was in short supply. Government work relief, according to Bean, wasn't "a very promising outlook for a man with any initiative." The traditional outlet for ambition in the plains was disappearing as larger landowners were incorporating nearby farms into their own. Bean described how wealthier farmers tore down the improvements made on smaller farms and cultivated their large acreage with power equipment. "Now what I want to know," he lamented, "is how in the devil can a poor man like me ever expect to get started in farming?"[14]

Aspiring owners in Kansas or anywhere else in the Great Plains found a culprit for soaring farm tenancy rates in the supposed "land hogs" who were devouring lesser farms. But an underlying cause was the farmer's need to achieve an economy of scale to make the farm profitable. The model for farming profitably in the plains during the 1930s was to grow, to spread operating costs across as many acres as possible. The alternative to increasing the size of the farm was to change the operation, farm one's cropland or pastureland more intensively, or settle for a lower standard of living than desired. Farmers in the eastern plains were at a disadvantage, despite falling land prices in the 1920s and 1930s, because they were less able to increase their production through more acreage. Wetter periods in the eastern plains produced lower yield increases than in the western plains. Compared to the years during and after World War II, the eastern plains experienced relatively little consolidation of farms in terms of larger farms snapping up smaller ones. Between 1930 and 1935, for example, most growth in farm size was limited to scattered counties in the western plains. During World War II, however, the size of farms increased in nearly all plains counties, save for a few in the eastern reaches.[15]

Farm observers saw another cause for higher rates of farm tenancy after World War I—land prices that were high relative to their income potential. This was an extremely volatile time for land prices in the Dakotas, Kansas, and Nebraska. After their zenith in 1920 at $66 an acre, farmland prices plummeted until 1940, when the price of an acre shrank to $17 an acre, a 75 percent drop from its 1920 value.[16] While land prices plummeted, so did farm income. Therefore, most plains tenants could not afford to purchase land, even at its reduced cost.

Another reason for higher tenancy rates was higher taxes on farm property. In 1937 one representative of the Nebraska Farm Bureau from the western part of the state made the dubious claim that his neighbors who stayed off relief did so by living in sod houses, thereby lowering their tax valuations.[17] Still, property assessments had skyrocketed over the decades. The tax on an acre of plains farmland nearly doubled from 16 cents before World War I to 29 cents an acre during the 1930s.[18] The boost in assessments was to pay for new and expanded services, mainly better roads and schools for rural residents.

One plains native examined higher tenancy rates in depth. In a study published in 1941, economist Robert Diller investigated Diller Township in Jefferson County in southeast Nebraska. The township was named after one of Diller's ancestors. Diller denied that concentrated landownership or absentee ownership increased farm tenancy. Rather, he saw it as a result of a combination of factors. Before World War I, lower land values, higher farm income, and good weather accompanied by good yields enabled many tenants to buy their land with a five-year mortgage. However, after the war these conditions disappeared. During the twenties and thirties young farmers in Diller Township chose or were obliged to rent their land rather than purchase it. This suited retired settlers well since they preferred to lease their homesteads rather than sell. Diller concluded that retirees or inheritors owned 43 percent of all rented land, and thus farm tenancy was a "natural" accommodation to the changing farm climate of the 1930s. Many of these landlords acquired the land themselves from relatives. Diller estimated that 60 percent of landowners in the Nebraska township inherited their land. The remainder was purchased. In Diller's opinion, only 10 to 20 percent of land in the township was not part of this intergenerational exchange of land. This "abnormal" land passed into tenancy because of drought, economic depression, and foreclosure. During the relatively prosperous two decades before World War I, this land would have been purchased.[19] Furthermore, Diller saw improving conditions in his home county. Since settlement, land tenure had become more stable. Landowners occupied their property longer and the proportion of land exchanged by inheritance rather than by sale had increased. Furthermore, the distribution of land in the township had evolved from half the land owned by a few absentee large owners to widespread ownership among many owners of different sizes.[20]

Robert Diller made valid points. Since the frontier days, landownership in the plains had probably become more stable. Furthermore, a substantial proportion, 28 percent of plains tenants and 19 percent of American tenants, were related to their landlord and would probably pass into ownership.[21] There was certainly a substantial class who eventually inherited their land. However, Diller failed to address significant issues beyond the relationship between landlord and tenant and the stability and disposition of landownership. The critical questions of the 1930s dealt with how tenants were able to run their operations, to maintain their standard of living, and how they treated the soil. Therefore, the agricultural, economic, social, and environmental attributes of plains tenant farms demand investigation.

One Farm, Two Farmers

For the lower tier of renters, the primary problem with farm tenancy was that two farmers, the owner and the renter, made their living off the same land. Also, during the 1930s critics complained that plains tenancy meant farms with low incomes and deteriorating soils, buildings, and fences. Furthermore, many believed that renters were unable to increase their livestock and their general productivity. One way to understand plains renters more fully is to compare them with farm owners (see table 1).

From a national perspective, plains tenants appeared better off than their cohorts across the nation. Table 1 demonstrates that, compared with average American farm tenants, those in the plains had larger farms and were more mechanized. Owning their own tools and livestock, most plains renters were certainly not as impoverished as black and white share-croppers in the Cotton Belt. Compared to plains farm owners, tenants in the region appear better placed to weather drought and economic depres-

Table 1. Comparisons of Average Full Farm Owners with Average Farm Tenants and Their Operations in the United States and the Great Plains, 1940

	U.S. Farm Owners	U.S. Farm Tenants	Plains Farm Owners	Plains Farm Tenants
Age	53 years	42 years	55 years	41 years
Year of occupancy	1923	1934	1920	1933
Farm size	124 acres	132 acres	280 acres	349 acres
Farm value (buildings and land)	$4,960	$4,569	$7,243	$6,900
Value of all farm buildings	$1,978	$1,215	$2,297	$1,602
Value of all machinery	$560	$499	$835	$862
Farmers working off farm and number of days	30% (166 days)	27% (106 days)	20% (120 days)	30% (71 days)
Farms with tractors (with average model year)	24% (1932)	21% (1933)	44% (1932)	54% (1932)
Farms with automobiles (with average model year)	64% (1933)	49% (1932)	82% (1932)	84% (1931)
Farms with telephones	29%	20%	45%	31%
Dwellings lit by electricity	42%	16%	29%	15%

SOURCES: Bureau of the Census, *Sixteenth Census of the United States: 1940, Agriculture*, vol. 1. *First and Second Series: State Reports*, pt. 2: *Statistics for Counties* (Washington DC: GPO, 1942), state tables, 1, 2, 9–11; Bureau of the Census, *Sixteenth Census of the United States: 1940, Agriculture*, vol. 3: *General Report, Statistics by Subject* (Washington DC: GPO, 1943), 148, 326, 359, 393, 452–56.

sion since they had more acreage and lower machinery costs per acre. In other respects, the plains owner and the renter were similar. The value of their farm and farm machinery were roughly alike, and both were more likely to have a car in their yard.

On the other hand, plains farm owners were better off than farm tenants in critical areas. While the tenant had more acreage, the value of the owner's property (including farmland and buildings) was 24 percent higher per acre. The value of the owner's farm buildings was 30 percent more than that of the tenant. This supports the public's image of poorer land and rundown buildings for renters.

When comparing tenants with owner-farmers, agricultural experts considered the impact of tenancy on renters' cropping and livestock operations, their use of machinery, and the likelihood of soil erosion on their lands. Only one study offers comparative data on plains owners' and tenants' use of land and livestock practices. In 1939 the Nebraska Agricultural Experiment Station researched full-owner and tenant farms in Box Butte County, Nebraska, in the ranch area in the state's panhandle. By indexing the productivity of the two classes of farms, researchers found differences between the two. First, there was a decided difference in land use. Full owners split their land roughly in half between pasture and crops. Tenants, however, used two-thirds of their land for crops and one-third for pasture. They also tended to plant more of their cropland in small grains for sale on the markets than did full owners. Forty percent of the surveyed renters were dissatisfied with this arrangement. They would have preferred to take some of the land devoted to cash grains such as wheat and transform it to feed and soil-building crops, such as alfalfa, or turn that land into pasture.[22] However, as tenants they had to abide by their landlord's lease, which often assigned acreage to marketable grains like wheat.

Renter-farmers also had fewer livestock. The Nebraska report gauged this by comparing full-owner and renter "livestock units." One horse, one cow, five hogs, seven sheep, forty turkeys, or 100 chickens are each equal to one livestock unit. In Box Butte County, tenants had 46 percent fewer head on an average farm than did farm owners. Theoretically, since they were free of mortgages, tenants should have invested more in livestock than did owner-operators. However, low income, high mobility, and the poor quality of barns and pasture on rented land made profitable animal husbandry difficult. Although tenants may have had adequate buildings or pastures on their current properties, they could not count on them in future leases. Indeed, Box Butte County tenant-operators estimated less than half of their livestock could be satisfactorily housed. Many surveyed tenants wanted to raise more livestock.[23] Again, they did not, probably because their landlords were more concerned with increasing their own income than with building up the soil or diversifying their tenants' operations. Plains landlords themselves were strapped for income and wanted as much from their rental properties as possible to pay off their own debts and expenses. Growing cash grains such as wheat was the accepted moneymaker in the plains.

The renters' relative overdependence on cash crops and their inability to increase livestock production suggests a weakness of tenant agriculture. It probably hurt tenants' opportunity for domestic self-sufficiency. The Box Butte County study suggests full farm owners were more diversified in their land use, less dependent on potentially soil-depleting small grains, and better able to maintain and increase their livestock production, compared to tenants. Therefore, owners had a long-term capacity for building up their soil and their livestock.[24]

The real advantage in tenant farming during the 1930s should have been in mechanization. Historian James Malin cited farm tenancy as a method of avoiding risk in the plains, where conditions changed so often. Through renting, according to Malin, the landowner (the landlord) split the risks with the machine owner (the renter). Malin saw tenancy as "a necessary adjustment to commercial agriculture in a machine age with high capital requirements."[25] In 1940, as noted in table 1, plains farm tenants were more likely to have tractors, higher values of machinery, and an equal proportion of cars compared to full farm owners. Since they could invest their capital in more rented land rather than in a farm mortgage, plains tenants spread the cost of their machinery expenses across larger fields to make an average investment of $2.47 per acre. This was 18 percent less than farm owners invested in machinery per acre.

Perhaps the most persistent charge against farm tenancy was that it damaged the soil. During the 1920s and 1930s renters were probably tempted to push their fields to the limits of their production. Anecdotal evidence between the two world wars suggests that farm tenants, with their high capital costs, unstable commodity markets, and unsure tenure, saw little need to conserve the soil. In spring 1916 a Kansas farm tenant wrote

Country Gentlemen about "Penurious Landlords and Prodigal Renters." He noted that his landlord, "Sam Bolton," did little to improve soil fertility on his land. The renter observed that the land's yield of wheat per acre had been dropping from 25 to 20 to 17 bushels in recent years. "So," wrote the tenant, "Sam and I are pals in robbing the soil. . . . Just what the future holds in store for us renters none of us is able to foretell. It's pretty hard to feel that we really have a grip on the future, for maybe somebody else will be farming this land next year. So we are going to squeeze out of it every dollar we can while we're here. And we expect to do the same to the next place we move to." The renter concluded, "The only way Sam Bolton will ever return to the soil will be in a nice long box with silver handles."[26]

There are no data to compare soil conservation practices between farm owners and renters during this period. In 1926 the USDA noted soil depletion in rented farms in the Dakotas and Kansas, where one-third of all rented farms showed decreasing soil fertility.[27] Others made firsthand observations about farm tenants and their land use. J. H. Beckmann, a Lutheran minister in Albion, Nebraska, wrote in 1935 that without ownership of the land, the tenant "farms the soil to death. . . . Ditches and gullies wash deeper each year, trees are not planted, pastures are overworked. He hires no help and farms every inch that looks like soil so that he may raise a quick crop."[28] Experts such as Cal Ward, the plains regional head of the Resettlement Administration, agreed. He noted that because of uncertain tenure, the renter ignored soil conservation. Ward argued that the tenant neglected erosion control, crop rotation, or commercial fertilizer to build up soil when he may have to leave within a year.[29]

The general perception of the era equated farm tenancy with soil depletion. The evidence is based on direct observation and cannot be ignored; however, it is not comparative and has little data supporting it. Surely the renter had plenty of motives to farm the soil for all it was worth, then move on without restoring it. But where is the evidence to conclude that renters caused soil erosion? William Harbaugh has reviewed testimony similar to that cited above, found it wanting, and noted "hard evidence that tenants 'mined' the soil more intensively than owners is surprisingly thin." He admits, however, that tenants planted fewer soil-building crops than did owners. Furthermore, many landlords pressured their tenants to squeeze as much as they could out of the fields to increase their profit from the harvest.[30]

Renters under short-term leases were under considerable pressure to raise soil-depleting cash crops on as much land as possible. However, many farm owners surely damaged their lands' fertility through harmful cultivation as well. Available evidence reflects the complex nature of tenancy in the Great Plains. On one hand, the 1935–36 National Resources Planning Board (NRPB) study found plains farm owners and tenants had comparable incomes.[31] On the other hand, there were many signs of near-indigence among plains renters. A 1935 study of eleven hundred farm owners and tenants throughout South Dakota found many farm tenants had only a pretension of security, since their total income was nearly one-quarter less than that of farm owners. Both groups had scant reserves or investments outside the farm to carry them through the hard times of the drought- and depression-ridden 1930s. However, farm tenants were especially ill-suited to meet the twin challenges of surviving hard times while increasing the scale of operations. Plains farmers, like most farmers in the past, walked a tightrope between using cash income to increase their standard of living and using it to invest back in the farm. Yet the low total income put extreme financial constraints on family and farm spending. According to the South Dakota figures, tenant farmers spent half their income on family expenses, which was still 20 percent less than the household spending of the farm owners down the road. Farm owners and renters reinvested close to the same percentage of their income into the farm operations. However, tenants had only a third of their income remaining after subsistence to invest in their business. But perhaps it made little sense to invest in a farm operation that was only temporarily theirs.[32] Government investigators also found that tenants were three times more likely than farm owners to resort to government assistance. While 8 percent of NRPB– surveyed farm owners in Kansas and North Dakota were on relief in 1935–36, one-quarter of all tenant farmers in the study received government aid.[33]

The Farm Tenant and Mobility

Apologists for farm tenancy in the 1920s and 1930s suggested that farm tenancy was the tool young men and women used to enter agriculture. They saw tenancy as a stage in a career where beginning entrepreneurs advanced up the steps of the occupational ladder to farm ownership. Therefore, leasing one's land held not only agricultural and economic considerations but social considerations, too. Two issues arose when discussing the lease arrangements between landlord and tenant: the method of renting the farm and the length of the lease. Tenants in the plains in 1940 used three types of payment: (1) cash payment, which was the primary kind of arrangement in the ranching area of the Nebraska Sandhills and western South Dakota, (2) share payment, which split the crop between the landlord and the tenant, a method that was popular in western Kansas, western Nebraska, and the northern half of North Dakota, and (3) a combination of share and cash rents, which dominated the eastern parts of Kansas, Nebraska, and South Dakota, and the southern portion of North Dakota.[34]

The terms of payment varied from farm to farm, based on tradition, force of will, farming conditions, and market demands. The split of crops between the landlord and tenant, for example, ranged from a fifty-fifty split to a two-thirds share to a three-quarters share for the tenant. Many leases were simply verbal agreements acknowledged over a handshake. For a plains farm tenant, the one- or two-year lease on a property offered either hardships or freedom, depending on one's point of view. Five hundred farmers meeting in Salina, Kansas, clearly saw the short-term lease as a detriment. In a petition to President Roosevelt and others, the farmers called for longer-term leases to permit "the principles of conserving soil fertility and maintaining a balance between production and market requirements." The short-term leases, said the farmers, had been a "stumbling block" for such improvements.[35] The Resettlement Administration as well asserted that year-to-year leases seriously hampered tenants from organizing and running their operations for long-term improvements.[36]

Others, while not praising the one-year lease, saw its utility. Robert Diller, in his study of land use in southeast Nebraska, concluded that the short-term lease was not as bad as it seemed. Such agreements left both parties an effective way to respond to market and other important changes. Diller noted that half the time tenants moved off a farm to acquire their own land, to rent a better farm, or to leave farming entirely. Landlords ended one-third of the leases because they wanted a better farmer on the land. The rest involved landlords selling the land or they themselves or a relative moving onto the land. Diller argued that tenant farmers of character received longer-term leases, and he suggested that once tenants stayed on their property for more than three years, they settled there permanently.[37]

Many renters actually preferred short-term tenancy and the freedom it offered. Harold Clingerman was a manager of farms in central Nebraska that his bank had gained through foreclosures. In 1940 Clingerman was constantly prodding tenants to buy foreclosed land, but with little success. One renter replied that he would like to, but "I don't even know if I can stick it out another year" due to marital problems.[38] Many renters understandably refused to get sucked into a bank mortgage on farmland in a depressed economy. The option to leave their rented property was the only freedom they had.

Moving day for most plains farm tenants was March 1. By that time renters had fed all their own livestock fodder to their cattle, and they preferred to move into their new farms to get settled in before field work began in April. Certainly most renters did not move every spring, but they were much more mobile than farm owners. As table 1 points out, in 1940 full farm owners in the plains had been on their current farms for an average of twenty-nine years, while tenants had been on their land an average of only seven years. Particular areas of the plains had a high annual turnover rate for tenant farmers. During the middle 1930s the eastern quarter of Kansas and scattered counties in western Nebraska and the western Dakotas reported 40 percent or more of their renters had operated their farms for two years or less.[39]

One result of this high mobility was poor housing conditions for many plains tenants. Farm tenant families moving frequently found their homes more rundown than those owned by their neighbors. So, despite their larger acreages, tenant farms were worth less, particularly in buildings, than owner-occupied farms. Surveyed Kansas and North Dakota tenant farm homes were worth only $1,024, or a third less than the farmhouses of owners, which averaged $1,552 in value.[40] Many tenant homes and farm buildings stood out because of their lack of repair. Travelers along country roads could distinguish a tenant farmstead from an owner's by the former's shabby appearance. Again, landlords saw little incentive to keep up the farmhouse and outbuildings for tenants. For example, they may have seen no point in renovating a rented farm home for electricity or phone service when the next tenant might not be able to afford utility bills. In fact, plains tenants were much less likely than owners to have electricity or phone service in their homes (see table 1).

Landlords had personal reasons for not investing in their rental properties. One Nebraska landlord explained the poor state of improvements on a farm he rented: "I spend as little money and do as little as possible in the way of repairs, for three reasons. First, whatever I might do, the tenants would ruin it; second, I carry no insurance. . . . I want to keep my possible loss down to a minimum; third, I don't give a damn."[41]

Similarly, Harold Clingerman found he could expect little from some tenants. He wrote terms into his leases requiring renters to cut weeds and maintain buildings and fences on the properties he managed, but he had little recourse if they ignored these terms. On one visit to a tenant's kitchen, the bank agent found rickety furniture, ashes spilling out from the coal stove, and a filthy oil cloth covering the kitchen table. Incidentally, Clingerman found that the locals disapproved when he discontinued long-term renters on bank-owned lands, even when he found a buyer for the foreclosed property.[42]

Many farm tenant families may have looked at their rustic living conditions as the price paid for running their own businesses, a kind of "deferred gratification." If the tenant family's farmyard and home were run down, they had more to invest in a tractor or Chevrolet. If they lived without a telephone or electric lighting, so did the most farm owners before World War II. Apologists for tenants contended that the lower-valued farm buildings and drop in home amenities were a sign of entrepreneurs just starting out. Renters were younger and occupied their farms fairly recently. Like many young businessmen and women, they ran cheaper operations and gave up a higher standard of living in hopes for a better life in the future.

Agricultural, economic, and social data offers compelling evidence on farm tenancy, but personal testimonies put flesh on the statistics. Families in Kansas, Nebraska, and South Dakota voiced the frustrations and triumphs of farm tenancy. They show the gradations from barely making do to a moderately comfortable life. R. H. Burch of Haviland, Kansas, wrote of his unhappiness as a farm tenant. In 1938 he had been renting a 480-acre farm for ten years. Since 1928 he paid his rent with a combination of one-third of his crops and cash, which totaled $3,000 over the decade. In addition, he made repairs on the place and wore out another $3,000 in machinery working the land. After subtracting his high operating costs, the Kansas tenant probably lost money from a decade of farming. For his troubles, Burch reported, a large landowner bought the land, and now Burch was living in town and running a farm half the size of his old one.[43]

In her study of Boone County, Nebraska, farm women, Deborah Fink interviewed "Beth," a farm tenant, about her experiences with her husband, "George." Between 1930, when they were married, and 1940, George held no fewer than ten jobs and Beth had four different jobs. Throughout the decade, the couple entered and left tenant farming according to their needs and resources. In 1932 they rented an eighty-acre operation. During their first year, which Beth remembered as the hardest of their marriage, their only cash income was from cream Beth sold in town. They lived on a diet of wild greens, potatoes, milk, and garden vegetables. In 1933 Beth and George rented a larger, 180-acre tenant farm. Despite some good crops and government aid, the couple's farm failed in the face of low commodity prices, drought, and grasshoppers. So, in 1935 they sold their farm equipment and moved into town. For the remainder of the 1930s George ran a service station, then moved the family to Wyoming where he worked as a ranch foreman. Beth had to perform housekeeping work for his crew. In 1940 the couple returned to Nebraska to live with Beth's parents, while George worked a 160-acre farm.[44]

Another example, that of Gladys and Ray Gist of South Dakota, paints a portrait of a farm tenant family that eventually achieved stability through conservative spending and the neighbors' support. Married in 1920, the Gists lived on nine different farms until they finally settled near Madison, South Dakota. Unlike many tenants, the Gists were slow to mechanize. During the 1930s they had a tractor and an automobile—but neither ran well. Only in 1941 did Ray buy a used Farmall tractor. He also shared farm implements with neighbors and hired threshing crews during harvest. Anyone who doubted Gladys's thrift had only to see her gas-powered Maytag washing machine—which she used for twenty-five years. Many complained that tenants, because of their high mobility, rarely took part in rural social affairs. However, the Gists were quite active in community life, wherever they lived. Whenever they moved, they joined a local church and participated in social activities in the rural neighborhood. These community ties saved them, for when their rented farmhouse burned to the ground, their church and local merchants gave them money. By the early 1950s the Gists had repaid their debts and enjoyed a middle-class lifestyle in their middle age.[45]

All the evidence surrounding plains farm tenancy during the 1930s supports several conclusions. Renting one's land offered numerous advantages over landownership. Leasing farm property allowed tenants to work more land, purchase more agricultural implements, and lower farm expenses per acre. However, many tenants lived in poorer quality homes and their farm and livestock buildings were inadequate for expanding production. They probably were overdependent on small-grain agriculture, which kept them from diversifying into a more stable regimen of livestock and feed crops for the time. Although plains tenants avoided a mortgage and property taxes, they had less financial security.

This evidence depicts a class within the rural plains that was on the margin between indigence and security. Not surprisingly, plains tenants were much more likely than owners to depend on government relief and the rural rehabilitation program to survive.

Unfortunately, the environmental and economic hardships in the plains during the 1930s made profiting from the land difficult regardless of tenure. However, federal and state government policies favored farm owners over tenants. For example, the New Deal's Farm Credit Administration allowed distressed landowners to refinance their mortgages. Local and state governments offered property tax relief for farm owners. In contrast, the depression decade was especially brutal on many renters, and the social costs of farm tenancy must have seemed exorbitant. During a decade of monumental unemployment, drought, and inadequate relief measures, the benefits of tenant mobility were questionable. Finally, despite poor farming conditions, the glut of tenants seeking land made it difficult for them to improve their situations.

Clearly, plains tenant farmers were geographically mobile, but did this translate into upward social mobility leading to landownership? Jeremy Atack suggests that from the 1860s to 1930 American farm tenants ascended the ladder through their life cycle . He sees the years of toil by younger tenants as part of a cumulative process in which they gathered the capital and expertise to progress from laborer to tenant to part-owner to full farm owner. By the 1930s, according to Atack, the rate of younger farmers moving from tenancy to ownership or leaving farming actually increased faster than in previous decades.[46] Historian Allan Bogue also contends that by 1930 the percentage of farm operators who rented their land plummeted as they grew older.[47]

Certainly, some plains farm tenants were able to acquire their own farms during the 1930s and 1940s. There were scattered examples of renters becoming owners. In late 1936 the Federal Land Bank of Omaha noted that tenants bought a quarter of the farms the bank sold that year. A bank treasurer encouraging downcast farmers said that "now would be the worst time to quit. . . . You would be letting go at a period when land is selling at its low point. Those who stayed on their farms during other lean times saw their land increase greatly in value, and I think those times will come again."[48] In addition, individual states had policies enabling tenants to buy land. In North Dakota, the state's Anticorporation Farming Law of 1932 compelled farm corporations to sell their lands within the next decade. This profusion of land on the market held down land prices and allowed renters to buy land during the wartime recovery of the early 1940s.

However, the 1930s were not an easy time for the farm renter to become an owner. The environmental and economic conditions of the decade made keeping a farm, much less buying one, very difficult for borderline farmers. No census material of the 1930s is available to verify whether tenants were truly becoming owners. The research of Atack and Bogue suggests that the older the farmer, the more unlikely he would be a tenant. However, the lower number of older tenants perhaps means that younger ones became frustrated and left farming before they reached middle age.

One survey paints a portrait of a tenant class without a ladder into ownership. During summer 1935 the federal government polled plains farmers on relief about their usual and current occupations. Three-quarters of those who were normally tenant farmers remained so. Twenty-one percent of renters described themselves as unemployed and looking for work, the remaining few had become farm owners, agricultural laborers, or worked in nonagricultural jobs. If few renters used the exit door, fewer farm workers or farm owners entered into tenancy. Only 3 percent of those who were usually either farm owners or farm laborers had become farm tenants.[49]

The evidence suggests that the lower tier of tenant farmers, or at least those on relief, had little choice between renting their farms and unemployment. Evidence also suggests that the increased proportion of tenant farms among total farmers came from men and women living, at least temporarily, off the farm. Perhaps many farm families had experiences like those of Beth and George of Boone County, Nebraska, who were raised on farms but moved between farm and town jobs, searching for some sort of economic stability. If so, the data and testimony imply that tenancy for most plains renters did not mean upward social mobility during the Great

Depression. Rather, tenancy often meant the farm family was just a step away from poverty.

Spotlighting the "Dispossessed"

Concern over the increase in farm tenancy rates and the widespread belief that tenancy led to low income, depressed living standards, social instability, and soil erosion made it a national issue during the Great Depression. Americans were well acquainted with the problems of the Cotton Belt. In the Deep South low cotton prices and long-term poverty dealt cruel blows to both white and black sharecroppers. Furthermore, southern cotton planters took advantage of the New Deal crop support program to defraud and displace their tenants. Rather than splitting commodity payments with their tenants, landlords pocketed the checks and kicked them off the land. The Southern Tenant Farmer's Union was formed in Searcy, Arkansas, in 1934 to champion the plight of dirt-poor southern sharecroppers. This tenant group's chief accomplishment was to expose the wretched working and living conditions of southern tenants and the planter's fraudulent and socially disruptive involvement in the New Deal cotton support program. The revelation of southern tenants' poverty and poor farming conditions also contributed to the passage of the Bankhead-Jones Farm Tenancy Act of 1937, which established the Farm Security Administration, the successor to the Resettlement Administration.

Since the 1920s at the national level, farm tenancy became synonymous with rural southern poverty and injustice, particularly among African Americans. However, Americans also viewed the Great Plains as a region of problem farm tenancy. What had changed during the 1930s was that voters, the press, lawmakers, and New Dealers were willing to invest federal power and resources to investigate and arrest the decline of farmland ownership. During the thirties Nebraskans shared their apprehensions. The Lutheran minister from Albion believed that solving the tenancy issue would correct many rural problems. Reverend Beckmann wrote President Roosevelt in 1935 that with low-interest loans the tenant "would have a chance to catch up on better power and machinery, and could believe again that someday he might own the land he was working, [and] the farm problem would take care of itself. The government would not have to pay such farmers to rest their land. Over-production would vanish." The minister warned that the renter was "working his land to death, staking even the life of his soil against more bushels because he has been tied down to a note or mortgage which is coming due."[50]

The *Lincoln Star* bemoaned that the older farmers who had lost their farms through foreclosure couldn't retrieve them without government aid.[51] The head of extension for Nebraska's agricultural college was vexed about the economics and climate of the era which made landownership less appealing. "It is true that there are some farmers who desire to remain tenants and feel that they would far rather have the other man own the land than attempt to own it themselves," W. H. Brokaw admitted. "But we must remember that this is contrary to the American ideal," he implored. "I'm sure that I need say little about the ambition to own a home which has been the safeguard of America."[52]

Republicans as well as Democrats railed against farm tenancy and its impact on American rural life. Governor Alf Landon of Kansas spoke on farm tenancy during the presidential campaign of 1936. In Oklahoma City, he called farm tenancy "one of the most serious long-time problems confronting the nation." Landon declared that "the stability of civilization depends upon ownership of the land by the man who works the land. The owner-operated farm is the foundation of sound agriculture." Saving the family-owned farm meant no less than "preserving individual opportunity," according to the Republican presidential nominee.[53] Secretary of Agriculture Henry A. Wallace concurred. After the election, he spoke on the problems of rural poverty throughout America. This indigence, said Wallace, led to "a corrosion of our rural life at its very roots." He attributed rural destitution to four causes: depleted soils, drought, farming in semi-arid areas, and the insecurity of many farm renters.[54]

The issue of farm tenancy was a cause célèbre before World War II. The yearning for security in the soil touched the hearts of Americans during the turbulent Great Depression. Best-selling novels of the 1930s such as *God's Little Acre*, *The Grapes of Wrath*, and even *Gone with the Wind* depicted people struggling for a home on the land.[55] Conservative and liberal farm observers criticized the problem of landless farmers, each from their own perspective. The U.S. Chamber of Commerce saw tenancy as a function of the market system. Unstable commodity and farmland prices forced tenants to rent land that remained relatively expensive compared to the soil's earning power. The Chamber further pointed out the problems caused by short-term leases, insufficient income, and inadequate compensation for tenants who improved their rented property.[56] In contrast, writers such as Charles Morrow Wilson used an emotional appeal to raise sentiment against conditions on American rented farms. They were, in Wilson's words, "among the most rapacious and degenerate of the modern world; an overlapping disgrace to American government, courts, banking, to investment finance and miscellaneous ethics of ownership."[57]

Despite the national censure of farm tenancy after 1936, many close to plains agriculture saw the market as the solution. *Capper's Farmer* pointed out the benefits of farm tenancy and warned against wholesale efforts to convert the renter into an owner. The editor, Ray Yarnell, argued that most tenants advanced into ownership, provided they had the managerial ability and the persistence to stick with the farmstead during hard times. Unfortunately, not all farmers had these attributes, Yarnell wrote. "If you give 100 men a farm apiece, in 20 or 40 years some still will have their farms, others will have been dispossessed of the land." Farm tenancy to some degree, would always be present.[58] Robert Diller, in his study of Jefferson County, Nebraska, concurred. Farm tenants in his township did not need to be "reformed" since they were not a distinct social or economic class. Most tenants were like their owner-neighbors in ability and geographic stability. Diller saw the lease as a joint venture of sorts between the landlord and tenant. He regarded the major puzzle in tenancy as "the case of the missing purchaser" of land. The true problem was in "the proper disposition of family land."[59]

In America the promise of the land meant ownership of that land. However, farm tenancy flourished during the Great Depression. What made Americans notice it was that during the 1930s tenancy was associated with insolvency, lower income and living standards, and soil erosion. Furthermore, it threatened to become a permanent situation for millions of farm families across the nation. Particularly in the Great Plains, borderline farmers were locked out of their aspirations. Their relatively low income, low capital, and small acreages barred them from enlarging their scale of operations. This in turn kept them from advancing their income, opportunity, and security. As a result, borderline farm tenants failed to achieve economic security, much less enter the landowning class.

During the Great Depression most agreed that poverty and insecurity walked hand-in-hand with poorer tenant farms. In reaction, some questioned the property system that allowed such high rates of farm tenancy. Others responded that the market was the final, and the legitimate, arbiter for land tenure. They argued that the market, combined with individual ability, determined whether the tenant ascended the ladder to farm ownership. The next step is to investigate government intervention, which was meant to alleviate the impact of drought and economic depression in the plains borderline farm tenants and owners, first through relief, then through the rural rehabilitation and tenant purchase programs. The tortuous development of these programs began, surprisingly, under President Herbert Hoover.

3

The Development of Rural Rehabilitation

Politics, Harlotry, and Herbert Hoover

Times were bad in the early 1930s in the American Great Plains. The region's press officials, who in previous decades sought homesteaders to settle the land, lay bare its troubles. In spring 1931 M. W. Lusk, the treasurer for a string of South Dakota newspapers, wrote President Hoover, "The agricultural situation in this whole section is desperate. . . . South Dakota, and the states in this group, north, south and east of here, have rich soil and practically sure crops normally but prices are so low for what the farmers have to sell and are still so high for what they have to buy." Then Lusk pleaded, "There must be relief or a courageous facing of a new and lowered standard of living."[1]

During the settlement period through the 1920s in Kansas, Nebraska, and the Dakotas, newspapers trumpeted the great opportunities the plains offered to the farmer. However, these promises were deceptive, as farmers in the 1920s often kept afloat through large mortgages and operating debts, which they could not repay. Despite the fertility of the land and the explosive world agricultural commodity markets, the plains farm, for both environmental and economic reasons, fell short of the income needed for the growing standard of living of post–World War I America.

The Great Depression and the extensive drought of the 1930s struck all farmers but hit the mortgaged and often ill-adapted plains farmers especially hard. The economic and environmental forces of the thirties were cruelest to borderline farmers. These were small- to medium-sized rural entrepreneurs who grossed between $500 and $1,000 annually during the depression. Many rented rather than owned their land. As a group, they had moderate resources in terms of acreage and the value of their land and buildings. Not only had the short-term problems of the decade stymied these borderline farmers, but they had to face the agricultural and economic trends of post–World War I rural America that rewarded farmers highly endowed with credit, land, and farm machinery.

Under the adverse conditions that Lusk noted, many farmers of the Great Plains desperately needed assistance from the state and federal governments. Borderline farmers in particular found that crippling drought and depressed agricultural prices left them with little income with which to survive, much less maintain their operations. They were in an even poorer position to expand the scale of their production, essential in the changing countryside.

Despite the problems within the region, many in the plains were ambivalent concerning outside help, particularly when it came from the federal government. Initially, Americans, and westerners in particular, were ill-at-ease with the commitments that came with assistance from their government. However, when it became clear that the economic slump following the stock market crash was a deeply rooted economic depression, more Americans willingly accepted and grew to expect government assistance. The plains region suffered greatly between 1929 and 1936 before it accepted such aid. President Hoover, despite his efforts, never truly addressed rural America's problems during his term. It took President Roosevelt and his New Deal another two years to turn from general relief to a specific agency to rehabilitate the way borderline farmers worked and lived on their land. This chapter looks at the six-year journey to create the rural rehabilitation program under the Resettlement Administration. Later, the program would pass into the Farm Security Administration.

As agricultural income fell further and further during the early 1930s, American farmers should have had a friend in President Hoover. Seemingly, Herbert Clark Hoover was the right man in the right place at the right time for the disastrous bouts of economic depression and drought that pummeled the Plains. This son of the Midwest had experience in directing government agricultural policy, as well as relief and recovery operations. During World War I, as head of the United States Food Administration, Hoover effectively directed American agricultural production for its war needs. Between 1918 and 1922 he ran the American Relief Association, a humanitarian organization that distributed medicine and food to war-ravaged Europe and the Soviet Union. Hoover was a popular and well-respected man in much of America. He impressed Franklin D. Roosevelt of New York. "He is truly a wonder, and I wish we [could] make him President," Roosevelt wrote in 1920. "There couldn't be a better one."[2]

Herbert Hoover was an effective and adaptive administrator. As secretary of commerce between 1921 and 1928, Hoover progressed beyond the traditional role of his position to promote social welfare in America. During the economic slump of 1920–22, for example, he pushed for active government efforts for recovery. During his tenure in the cabinets of Presidents Harding and Coolidge, Hoover took a great interest in unemployment. Even though agricultural problems were formally outside his bailiwick, Hoover paid much attention to the state of American agriculture. Throughout the 1920s he viewed this sector as an unhealthy natural resource industry, like the oil, timber, and mining industries bedeviled by overproduction and shifting markets. In response, Hoover supported voluntary efforts to cut agricultural commodity surpluses. He also promoted the efficient distribution as well as increased domestic consumption of farm produce. Furthermore, Hoover promoted converting marginal croplands to pasture and diversifying areas committed to one-crop agriculture. What President Hoover desired was to bring the American farmer into line with the nation's modern industrial economy. Thus it was Hoover and USDA bureaucrats who prepared the ground for the New Deal farm program.

Hoover's considerable reputation and the booming urban economy helped elect him to the presidency in 1928. President Hoover was a capable administrator with broad interests, but his skills failed him in the face of a devastating global economic downturn demanding new solutions in the early 1930s. Though he was an active chief executive, Hoover had a narrow view on the role of government within America's economy and society. Hoover believed in a limited "associative state." He wanted to coordinate voluntary and cooperative action from the leaders of capital, labor, and agriculture. Hoover eschewed federal regulations and planning as impediments to individual freedom. His associative state was advanced compared to the political philosophies of his Republican predecessors, Presidents Harding and Coolidge. However, Hoover's circumscribed approach promoted cooperative action among industries whose self-interests were a higher priority than the nation's well-being. Furthermore, Hoover's top-down strategy meant that smaller businesses, farms, and labor rarely realized the benefits of government initiatives.

Initially, the American farm sector was numb to the massive impact of the economic depression. In summer 1929 America's leading economic indicators showed a decided slump. Residential construction, consumer spending, and employment each declined. On October 24, 1929, the stock market collapsed. Within one month, securities on the New York Stock Exchange lost $26 billion, or 40 percent of their value. Agricultural prices, however, remained relatively stable; six months after the crash, farm prices had dropped only 10 percent. However, beginning in spring 1930 farm commodity prices began a steady decline. At the lowest price level, in February 1933, the value of farm goods had dropped 62 percent since the stock market crash.[3] The American farmer labored in economic depression until World War II.

Though the press's opinions represented a broad cross-section of opinion, their views on government relief for the agricultural sector was not the lone voice of the countryside. The plains press was more conservative and less likely to represent Democratic and ethnic Catholic farmers in the region. A survey of the background of leading Kansas newspaper editors during the 1936 presidential election found they were overwhelmingly Republican and Protestant.[4] Aware of this, New Dealers solicited opinions at the grassroots level. They also sent their own investigators, such as Lorena Hickok, to the plains to assess conditions and voters' opinions.

Still, taken together, the American and Great Plains farm sector gave vivid commentary to both the 1930s economic depression and President Hoover's failure to effectively deal with it. By 1932 most farm leaders preferred one of two strategies to counter the farm depression. Along with Hoover, many supported the associational approach that sought a rational farm economy by organizing farmers into cooperatives. Ideally, within these institutions farmers and their representatives constructed credit systems and market controls as well as coordinated agricultural research and production to stabilize and increase their income. The second approach was to manipulate the American economy for the benefit of the farm sector. To this end, farm groups called for expensive subsidies and marketing controls that would enable farmers to garner higher income. By the beginning of the Great Depression two important farm groups in the Great Plains—the Farm Bureau and the Farmers Union—had their own plans to rescue the American farmer. Later, during World War II, the Farm Bureau and the Farmers Union battled over government activism in the countryside. Because of this, their backgrounds and outlooks are worth examining.

The Farm Bureau had a decisive role in implementing New Deal agricultural policy. It had the reputation as the upper-income farmers' group with ties to business interests. Begun in 1911 in Binghamton, New York, the first Farm Bureau was formed within the local Chamber of Commerce to educate farmers. By World War I an informal trilateral support group had emerged among county farm agents, state agricultural colleges, and county Farm Bureaus around the country. This alliance gave the Bureau a semilegal status with local USDA programs.

Financial, business, and governmental leaders helped create the unified national Farm Bureau in Chicago in 1919 to organize farmers on a national basis. Supporters of the Farm Bureau wanted a conservative and stable force in the countryside to counter the radical and labor unrest sweeping the country after World War I. These conservatives were particularly fearful that the Nonpartisan League, a farmer-based group that was promoting semisocialist measures in North Dakota, would spread to other farm states.

The Farm Bureau fostered capital-intensive, large-scale production of agricultural goods for sale on an expanding market. It also labored to increase commodity production and build cooperative marketing and private farm credit institutions for large individual farm operators. The Bureau consistently criticized organized labor and allied itself with business interests between the two world wars. At least in speeches the organization also opposed government intervention in the economy and saw the increased and more efficient marketing of American agricultural goods domestically and abroad as the true salvation of the American farmer.

The genius of the Farm Bureau was its unity, organization, and ability to latch onto government farm programs. Although the Bureau faced a potential split in its two bases of support, the Midwest and the South, it managed to compromise and become a united voice for substantial farmers. The Farm Bureau was a break from past agrarian groups in American history. When insurgent farmers formed the Populists during the hard times of the 1890s, they fought for equal economic opportunity with large financial interests. It was an idealistic battle against millers, meatpackers, railroads, and Wall Street to win southern and western farmers the opportunity they deserved. The Populists fought to weaken business interests that dominated the agricultural sector and made the agrarian a vassal to urban industrial powers. Forty years later, in contrast, the Farm Bureau attempted to actually become a powerful interest group and the preeminent broker of farm policies on behalf of larger commercial farmers. This was the beginning of the modern farm organization, which was to use traditional agrarian egalitarian arguments on behalf of middle- and upper-income farm families.

In 1932 the Farm Bureau favored the McNary-Haugen plan to create an Agricultural Credit Corporation that would purchase American farm surpluses and sell them on the world market. If the world price was lower than the domestic price, then processors, distributors, and retailers of agricultural commodities would make up the difference through an "equalization fee." American farm produce could then be "dumped" on the world market at cheaper prices.

At the beginning of the Great Depression, the Farmers Union simply wanted a guarantee that commodity prices would cover the cost of production. While these two plans might have increased farm income in the short term, neither truly addressed the problems that bedeviled plains farmers, particularly those walking the borderline between solvency and bankruptcy. The Farm Bureau's plan, though not the Farmers Union's, addressed commodity surpluses, but both groups ignored the effects of rural poverty and the increasing scale of farming on the countryside.

The Farmers Union began in 1902 in Rains County, Texas, where its secrecy and inexpensive membership dues appealed to poorer farmers. By the time of the Great Depression the organization's southern membership had declined and its membership shifted primarily to the north-central Midwest. Those looking to the national Farmers Union for a strong, ideologically consistent voice were disappointed. During the early 1930s, factionalism, a decentralized power structure, and its adherence to the outmoded cost-of-production plan weakened the group. In particular, the Farmers Union was split between those who favored farmers' cooperatives and supporters of legislative means to improve farmers' livelihoods.

Herbert Hoover wanted an affordable government farm program without the continual and widespread involvement of the government in farming, and he was not alone. Some of the leading voices in the plains opposed federal intervention before the stock market crash. Like Hoover, the *Kansas City Star* in 1928 opposed the McNary-Haugen plan. The farmer "doesn't want a government board to run his business for him," the paper stated. Instead, the government should help by sponsoring improved marketing and transportation facilities, promoting agricultural education, and fostering efficiency.[5]

Once elected, President Hoover preferred to remedy the nation's agricultural problems from the top down. His chief solution for falling agricultural prices was to take bulging supplies of grain and cotton off the market through the Federal Farm Board. Created in 1929, Hoover's Farm Board attempted to steady the erratic farm prices by purchasing American surpluses. Hoover approved such government intervention as a means of restructuring the farm economy. As planned, the Farm Board would first stabilize agricultural prices through acquiring surpluses. Then the federal government would help create cooperatives for farmers to market their own goods and provide their own credit. This, according to Hoover, would assimilate American farmers into the modern American economy without an ongoing commitment to permanent government programs, such as price supports. Begun in 1929, the Farm Board was supposed to be limited in scope and duration. However, President Hoover could not have foreseen the monumental impact of the global depression on agricultural prices. Soon after its creation, the Farm Board found itself frantically attempting to shore up commodity prices in the Great Plains after they began to drop in 1930. By summer 1932 the agency had purchased $90 million worth of grain in the Great Plains, which equaled nearly half the amount that was spent on stabilizing the nation's grain prices.[6]

President Hoover showed an unprecedented commitment to the American farmer. However, although the Farm Board spent millions to prop up plains farm prices, it was unpopular with farmers and conservatives throughout the region. First, this expensive program failed to halt the slide of grain prices. Second, some saw the Farm Board as a federal intrusion into the rural economy. For example, the *Omaha Daily Journal Stockman* acknowledged that farm relief in the United States had to progress beyond price stabilization. At the same time, the newspaper feared an American version of the "authoritarianism" of European and Soviet farm policy in 1930. In those countries, farm programs were "forced on the peasants regardless of their desires, on the theory that the good of the nation comes first."[7]

The plains farm sector was deeply divided over aid for individual farmers. One Hoover farm program—emergency seed loans to farmers in the plains—showed the ideological struggle Hoover and other Americans went through over government involvement in farm life. As a great drought settled over the American farm belt in 1930, state and federal land banks refused credit to farmers reaping scanty harvests. In response to the anguished cries of plains farmers, President Hoover approved loans for their market production—but not for their subsistence needs. Local committees made up of leading farmers and bankers approved the loans and county farm agents administered them. The program had major administrative problems, however. The loans were difficult to oversee, county agents resisted taking part in the process, and the program burdened many American farmers with more debt than they could repay.

Furthermore, by 1932 these loans were made under conditions especially difficult for plains borderline farmers to meet. For instance, applicants could not use loans to buy livestock or machinery or to pay off past taxes or debts. Also, the program denied loans to farmers who had not raised a crop in 1931 and farmers could not apply for credit to cultivate more acreage than they had the previous two years. This strategy was impractical in the Great Plains, where expanding agricultural operations and mechanizing could offset lower commodity prices.

Between 1930 and 1932 President Hoover confronted a recurring problem in the American West: the independent entrepreneurial farmer's constant appeal for government assistance. Despite their self-image as "rugged individuals," plains farmers, particularly those in the Dakotas, were accustomed to seeking government aid. Between 1921 and 1934 the western half of North Dakota and the northwestern quarter of South Dakota received seed loans in five to seven of those years.[8] The Great Depression hit these two states hardest in the plains region.

Unfortunately, Hoover lacked the political savvy to display his personal interest in the nation's welfare to the public at large, and to the American farmer in particular. As Kansas newspaper editor William Allen White wrote, "Politics is one of the minor branches of harlotry" and Hoover was frigid.[9] Hoover was caught between his strong ideological opposition to granting relief beyond strictly defined limits and his political concern for rural America.

President Hoover faced a dilemma between the need for government intercession in rural America and his view that the country must be left on its own to recover. In fairness to Hoover, he faced opposition to government intervention within his own administration and his party, as well as from the public. In 1932 Secretary of Agriculture Arthur Hyde complained that seed loans under the USDA "cannot by the furthest stretch be called good business." The federal government was lending "more money on thinner security and sustaining more losses than ever before in the history of money lending," he decried.[10] Clearly, these loans went beyond Hyde's idea of proper interaction between government and capitalism. Since the market had saved the nation before, it could do so again. Experience suggested that the country was merely going through another financial downturn that the it could ride out. After all, the nation rebounded from the 1920–21 economic slump ready for a decade of prosperity. In addition, between mid-1929 and March 1933, the United States went through no fewer than three partial economic spurts, each suggesting recovery.

For these and other reasons, conservative plains newspapers and individuals joined in a chorus of opposition to government aid for nonfarmers and farmers alike. In 1930 one South Dakota newspaper criticized New York governor Franklin D. Roosevelt for proposing loans to the unemployed in his state. Such a policy, the editorial fumed, would draw the indigent to New York state. "Those accustomed to 'bum' their way through life will naturally drift to where bumming comes easiest," the *Aberdeen News* predicted. "The prospect of starvation must be held out to certain classes before they will stir themselves." Instead, the newspaper pronounced, the government should subsidize industry to create jobs.[11] The *Kansas City Star* also opposed federal involvement in local relief. In late 1931 the newspaper insisted that local communities should care for their own, even during a devastating economic depression. "The dangers of direct federal relief now being advocated . . . are so great that it ought to be avoided," the *Star* warned. "Once the precedent is established of looking to Washington for aid, the door is thrown open to all sorts of extravagancies."[12] Likewise, a small Kansas farmer vented a free-flowing diatribe on the "bum farmers" he saw. Writing President Hoover before the 1932 election, F. W. Morse claimed that he was getting along quite well on only four acres. What of the "bum farmer" receiving government aid? Morse declared, "Why he is over in the park, sitting there from morning to night cussing Hoover and eating Red Cross flour and waiting for Mr. Hoover to put several million dollars in the bank so he can borrow it on poor security or no security and never pay it back." "Bum farmers," Morse complained, "won[']t work and the Government is going to feed them [. . .] isn[']t that a fine state of affairs[?] . . . The more you help people, the more helpless they become.[13]

At every step of the way Herbert Hoover faced opposition from the nation, the Great Plains, and his own conscience in providing direct aid to farmers struck by drought and the economic slump of the early 1930s. Many saw relief and government involvement in the countryside as detrimental to farmers in the long run. Therefore, initially President Hoover had only mixed support for offering relief. Before they realized the severity of the economic slump, influential plains newspapers looked to market forces, self-sufficiency, and local relief to correct the region's insolvency and destitution. In addition, Hoover believed it intrusive, expensive, and politically unwise for the federal government to actively assist the American farmer at the grassroots level.

FDR and Cries for Help

Farmers in North Dakota received campaign literature from that state's GOP central committee during the 1932 presidential election. "Vote Hoover," one flyer read. "Be safe." Voters in the Great Plains states of Kansas, Nebraska, and the Dakotas ignored this warning as they voted for Franklin Roosevelt over Hoover 63 percent to 35 percent, reversing the 1928 presidential election, when Hoover had won the region with 63 percent of the vote. The worsening depression and Hoover's ineffectual farm programs turned the rural vote against him. The message sent by plains farmers was not lost on a shrewd politician like president-elect Roosevelt, who knew that future rural support rested on his ability to aid the countryside and restore farmers' buying power.

In the 1932 and 1936 presidential elections Roosevelt made and followed through on promises appealing to the plains states, which he carried in both elections. Over half of all plains representatives sent to Congress in 1932 and 1934 were Democrats. North Dakota sent Progressive Republicans William Lemke and Usher Burdick, who also supported the New Deal.

Roosevelt's election and his supporters in Congress through the 1930s changed the nation. For Americans, and especially farmers, Roosevelt's New Deal established a new relationship with the federal government. The eminent agricultural historian Theodore Saloutos called the next eight years of the New Deal "the greatest innovative epoch in the history of American agriculture."[14] The New Deal farm program was novel because it attempted to use voluntary means, legislative power, and the federal bureaucracy to stimulate economic recovery. Robert McElvaine explains the New Deal as "cooperative individualism" replacing the "acquisitive individualism" of the 1920s. However, the New Deal was implemented through large, sometimes impersonal bureaucracies that displaced personal and local assistance and offended many in the rural plains.[15]

Under the Roosevelt administration, the federal government attempted for the first time to organize farm production, compensate farmers for soil conservation, and stabilize mortgage debt and farm income at the grassroots level through federal agencies. This contrasts with Herbert Hoover's ineffective attempts to spark a recovery by encouraging the farm sector's private leadership to cooperate on behalf of the nation's needs. The New Deal's farm goals were to stabilize the farm economy racked by a disastrous economic depression and drought, then to restore the countryside's purchasing power. For the first time, the federal government used agricultural regulations, commodity production controls, and government crop subsidies to speed the farm recovery. Yet it took three years before the Roosevelt administration pieced together a concerted program specifically for borderline farmers.

The administration initially acted to bolster the farm sector because both influential and minor representatives of the American countryside called for restoring the farmer's income. During the dismal winter of 1932–33, the head of the American Farm Bureau direly predicted, "Unless something is done for the American farmer we will have revolution in the countryside in less than twelve months."[16] Estella Beems, a middle-income farmer of Keya Paha County, Nebraska, seconded this call for government aid during the same winter. Beems, a sixty-eight-year-old widow with 320 acres of land, a $4,000 mortgage, and twenty-three years of farming experience, wrote a poignant letter to president-elect Roosevelt to urge action. She suggested taking money from the war debts that other nations owed the United States and applying it to the farmers' mortgages. "You would be surprised how quick prosperity would return," she suggested. Mrs. Beems pleaded urgently, "Help us before you help any one else . . . . The idle will find no work untill the Farmer is buying. It[']s a vicious circle. [H]ere in our county the farmers need all Kinds of machinery, harness and other equipment. They need building material, clothing, household furniture, why they need someth[ing] of every thing," she wrote. "But it tak[e]s money," Beems continued, "and money we have not. Will People in Washington see to the farmers now—not next year. But Now."[17]

Like many people within the plains farm sector, this farm woman believed in the preeminent importance of the farmer in the national economic recovery. Helping the Mrs. Beemses of the countryside was important for American legislators during the 1930s. Savvy politicians around the country knew they could not ignore farmers such as Mrs. Beems and remain in office. Throughout the decade, one-half of all House seats and nearly all Senate seats in America represented districts or states with substantial farm populations.

Perhaps the most important problem facing the New Deal farm program was the vast surplus of commodities. Hoover's agriculture secretary Arthur Hyde was at a loss for a solution in 1932. "I don't know what we are going to do with all this wheat and cotton," he admitted. "Nobody seems to want wheat and cotton . . . and everybody seems to be raising it."[18] The Roosevelt administration worked to alleviate this glut of produce through the most important initiatives to come out of the early New Deal, the Agricultural Adjustment Administration (AAA). It was New Dealers within the USDA, not farm organizations or state land-grant colleges, who put together this critical program. However, the New Dealers did depend upon the support of important farm groups such as the Farm Bureau to pass and implement agricultural subsidies.

Generally speaking, the AAA was a county-level program in which individual farmers signed "marketing agreements" with the USDA to cut their production for such important commodities as corn, wheat, milk, and hogs. They were compensated with government commodity checks. The goal was to reduce the nation's surplus and place agricultural prices on "parity" with the price of industrial products. Parity was a formula devised to bolster the farmer's buying power. Like so much of the legislation of the New Deal, the AAA was the result of a set of compromises. It was a concession to those in the farm sector who wanted to reform agricultural practices and those who wanted to raise commodity prices. It was also a compromise between those who desired crop restrictions and those who eschewed crop controls and favored marketing agreements, whose added costs fell to the consumer. Nearly all sectors of plains farmers took part in the AAA.

Henry A. Wallace, Roosevelt's secretary of agriculture, helped to create the AAA and decisively placed his imprint on the New Deal farm policy. The Iowa-born Wallace wore many hats. He was the son of Henry C. Wallace, the secretary of agriculture under Presidents Harding and Coolidge. The younger Wallace was the editor of the influential monthly, *Wallace's Farmer,* and also developed the first hybrid seed corn for use by farmers. Wallace believed that the countryside could use technology and management techniques to surmount rural poverty. Politically, Wallace was a Progressive Republican. Although he served under a Democratic president, he switched to the party only in 1936. As agriculture secretary, Wallace sought economic equilibrium between farmers and city dwellers. For this purpose, he supported the goals of the AAA and also promoted cutting farm debt, controlling inflation, and cultivating new foreign markets.

Along with attacking the huge commodity surpluses, the new president had to address the results of the drought and low agricultural prices that struck rural America in the early 1930s. In August 1934 President Roosevelt made a tour of drought-stricken central North Dakota. Along the road he saw signs referring to his support for the repeal of prohibition. They said, "You gave us beer, now give us water." Roosevelt commented, "That beer part was easy." During a speech in Devils Lake, North Dakota, Roosevelt confided after seeing the parched Dakota fields, "I cannot honestly say that my heart is happy today, because I have seen with my own eyes some of the things that I have been reading and hearing about for a year and more." Referring to the drought, Roosevelt said, "It is a problem. I would not try to fool you by saying we know the solution to it. We don't." However, he ended the speech with an encouraging note. "If it is possible for government to improve conditions in this state, the government will do it."[19]

At the national and regional level, two important farm groups reacted characteristically to the early New Deal agricultural programs. The Farm Bureau as a unified front positioned itself within the commodity supports program. Edward O'Neal of Alabama served as the national president of the Farm Bureau between 1931 and 1947. After O'Neal learned that presidential candidate Franklin D. Roosevelt favored a plan to cut agricultural surpluses by paying farmers to remove land from production, the farm leader endorsed it. By allying himself with the AAA, O'Neal presented himself as the nation's spokesman for all farmers. The Farm Bureau also played the godfather to the AAA and the county extension service by supporting local county agents. In 1933, faced by cuts from cash-strapped state and county governments throughout the plains, the county agent system received a second life by publicizing and administering local AAA programs. Other early New Deal programs such as seed and feed loans and soil conservation programs were run through the Bureau's allies, the county agents. Later in the 1930s, because of its organizational support of friendly New Deal programs, the Farm Bureau would demand a prominent voice in administering these farm programs as well as framing new farm legislation.

The Farmers Union, on the other hand, divided in leadership and policies, sent different messages to the Roosevelt administration. The farm group offered rhetoric and little unity during much of the New Deal era. Nailing down the Farmers Union's ideology nationally and in the plains during the Great Depression is a difficult task, since the organization in each state appeared to have its own tenets. However, state branches maintained a steady support for the family farm, cooperatives, pacifism, and a sense of being a disadvantaged class. Populism, particularly around the concepts of production and wealth, bound plains states' Farmers Unions in the later 1930s. Charley Talbott, president of the North Dakota Farmers Union, vehemently defended democratizing the farm economy. The problem for the plains farmer, Talbott wrote, was that the wealth of human labor and land had been amassed in the hands of "shrewd manipulators of the machinery of production." The solution was controlling commodity marketing through cooperatives. Not only must agricultural marketing change, but farm ownership as well. "We will never alleviate human suffering as long as we produce for profit instead of for use," Talbott wrote. The solution was the "cooperative and collective ownership of property" by the producers.[20]

Emil Loriks, head of the South Dakota Farmers Union between 1934 and 1938, echoed that message. During a 1937 convention speech recalling the populism of an earlier generation, he cried, "Shall we crucify mankind on the cross of profits, or can we save mankind through economic democracy?" Loriks saw the profit motive eroding society. In his eyes the desire for financial gain was not behind the Farmers Union's cooperatives. Rather, they were "a Christian and neighborly interest and concern for the common good and welfare of each member of society. Coops are the engines of democracy."[21]

The New Deal led voluntary programs for farmers to cooperatively raise their income and stabilize farming conditions. Logically, the New Deal's combination of individual initiative with collective organizing should have appealed to all Farmers Union members. However, many members feared the federal government's role. Despite past failure with the private farm cooperative system, they believed that the New Deal's "regimentation" was a greater threat than economic depression.

Several state branches of the Farmers Union were hostile to New Deal programs in the early years. For example, there were bitter divisions over the AAA. The commodity price support plan needed a private sponsor in the countryside. The Farm Bureau seized the role and the Farmers Union lost the opportunity to become the New Deal's favored farm group. Because of this, many Farmers Union members in the plains felt slighted by the original AAA program. Price supports overshadowed the Farmers Union's proposals to boost farmers' purchasing power. As a result, the Farmers Union charged the New Deal's AAA with favoring more substantial farmers in the Farm Bureau.

The Nebraska Farmers Union in particular attacked the New Deal with conservative populist rhetoric. After the 1936 Roosevelt election landslide the *Nebraska Union Farmer* insisted that Americans were approving New Deal assistance, not New Deal intervention into the economy. An "ordered economy," it stated, would lead to "nothing less than a profit-seeking dictatorship." Under such a system, "the people would be at the mercy of a rigid, cast-iron system, ostensibly controlled by the government, but actually controlled by the industrial masters."[22]

By the late 1930s, either by a change in ideology or through expediency, most Farmers Unions outside Nebraska supported the New Deal. During his 1937 "Cross of Profits" speech, Emil Loriks insisted, "We have the right to ask the federal government for aid and assistance in times of great emergencies. . . . It is the duty of government to come to the aid and assistance of any area that is so stricken."[23] That year, state heads of Farmers Unions in the plains met in Aberdeen, South Dakota, to discuss the drought situation in the region. After years of drought, the conference reported, farmers found themselves "unable to take care of actual human needs without immediate Federal assistance." The state Farmers Unions, including the Nebraska branch, then asked for liberalized farm grants, feed loans, subsistence grants, and expanded work-relief funding. The state leaders also requested an end to mortgage foreclosures on government loans when drought had cut the farmer's ability to pay.[24]

The Deserving and Undeserving Poor

For the first time in American history, the federal government committed itself to lifting great sectors of the countryside out of poverty and to reforming the agricultural practices of commercial farmers in the Great Plains. The agricultural, economic, and social problems of the 1930s were complex and obstinate. Added to these problems was the long-term "hard core" poverty among the aged, disabled, infirm, the children of female-headed households, new immigrants, and minority groups. Furthermore, outside of the economic depression there remained long-term trends in both urban and rural America that exacerbated the problems of the 1930s. These problems included an aging population, large annual increases in the labor force, unemployment remaining from the 1920s, vast soil erosion, and farm insolvency.

The most pressing problem for individuals in rural America was the cruel plunge in living conditions among many farm families. Unfortunately during the early 1930s, government assistance was very limited in rural areas. Historically, rural poverty was a private matter. Other than some state charity for the aged, the blind, and indigent children, no programs existed to promote the social welfare of the countryside, much less reform it. In rural areas county agents promoted efficient production among the more substantial commercial farmers. Home demonstration agents visited some counties to advance farm home economics. Yet these and a few scattered programs for rural health and hygiene were the extent of the social welfare force in the countryside.

American consciously created this limited public relief. The majority of Americans shared a unique set of convictions regarding the indigent as well as the solutions for poverty. Americans in the early twentieth century divided the poor into two groups: the "deserving poor" and the "paupers." Solutions to poverty were formed according to this dichotomy. For the "deserving poor," Americans believed that work was the paramount solution. As a result, public policy was formulated around the limited goal of preventing destitution for this class until they were back on their feet again, rather than maintaining or increasing their income.

For the paupers, or "undeserving poor," most Americans during the 1930s asserted that no one was entitled to a certain standard of living, even at subsistence level, without earning it. The public believed that these unworthy men and women either refused to work or lacked the resources to work themselves out of poverty. Therefore, providing them relief only made them dependent on the taxpayer and open to manipulation through partisan patronage. Many Americans saw these "paupers" who lived in American cities, towns, and the countryside as unable to escape their poverty. For these people, government aid beyond subsistence was not only undeserved but wasteful.

The plains reflected this dual reaction to government relief. On one hand, a strong faith in work and self-reliance survived in the Great Plains during the 1930s, despite widespread poverty and the lack of financial opportunities during the time. Perhaps this was because the frontier heritage valuing thrift, labor, and determination was still alive during the 1930s. Many residents of the countryside were only a generation or two removed from the hardships, toil, and deprivations the pioneers endured to convert the plains into productive farms and communities. On the other hand, thousands of plains farmers relied on New Deal relief programs to keep their families fed and clothed and their operations running.

These discrepancies biased the progression of New Deal goals from relief to actual rehabilitation of borderline farmers. Plains residents shared ambiguous notions about government aid. The region looked upon New Deal relief and rehabilitation programs with ambivalence, a mixture of gratitude and chagrin, that colored their administration. During the 1930s federal relief and rehabilitation programs were not part of some evolving acceptance of a liberal, activist role of the government in the city and countryside. Rather, at least in the plains, they were a genuine and necessary response to the drought and deprivation that Roosevelt himself saw during his visit to North Dakota in 1934. In other words, most plains residents accepted this New Deal activism out of short-term emergency and utilitarian needs, not out of a long-term shift in ideology.

The 1934 elections sent many left-of-center representatives to Congress. This pushed Roosevelt and his New Deal further into government activism. Furthermore, New Dealers may have seen these programs primarily as bulwarks against seemingly radical groups such as the Farm Holiday Association or individuals such as Huey Long and his Share-Our-Wealth scheme, which proposed heavily progressive taxes. In 1935 Long's appeal was surprising strong among midwestern farmers. Government officials believed that leaving the plains countryside to its own fate would only serve to prepare the soil for seeds of radicalism or irresponsible policies. Sober leaders inside and outside the region approved of relief for these limited reasons.

"You Reach Out and Take What You Can Get"

The administration of relief programs in the rural plains further revealed the complex attitudes the nation and the region held concerning government assistance. Before the establishment of the New Deal's Resettlement Administration in May 1935, American farmers received three types of aid from the government: (1) direct relief, or grants of money or supplies for subsistence, (2) work relief, which was employment on public work projects for which clients were paid for their labor, and (3) emergency relief, which in the plains often meant livestock feed and seed grants and loans in drought areas. The Roosevelt administration heard a deafening clamor for assistance coming from the heart of the nation. The Great Plains countryside continued in its crisis between 1933 and 1936, particularly in North and South Dakota where crop failures and low farm income threatened to turn borderline farmers into destitute farmers. In January 1934 the national headquarters for the Farmers National Grain Corporation reported serious conditions throughout the Dakotas. In counties throughout both states, farm families were short of flour, meat, coal, blankets, and clothing.[25] Ten months later, in October 1934, the situation was no better, according to Governor Tom Berry of South Dakota, who for nearly a year also served as his state's relief director. After successive years of crop failure caused by drought and grasshoppers, conditions throughout the state were degenerating in both the farm and the farm home. Berry related after a tour of South Dakota farms that "equipment used in farming operations are deteriorating to the extent that much of it is beyond repair; the household furniture is in very poor condition, barely holding together." Berry wrote that "articles of bedding such as mattresses, comforters, blankets, sheets, pillows and cases and many other articles of the home are practically worn out. . . . reserve articles of clothing and underwear have been used up." Many South Dakotans neglected their health, Governor Berry continued. Half the families in the state were on relief. Farmers used their subsidy and government emergency checks to repay loans, pay taxes, and buy feed for their livestock.[26] Probably little remained from commodity checks to maintain or enlarge farming operations.

From 1933 to 1935 the federal government created new agencies to assist those in poverty or on the brink of poverty. Rural families from all income levels demanded aid, and the New Deal set up agencies to meet the outcry for assistance. The most important agency for direct relief was the Federal Emergency Relief Administration (FERA), which ran between 1933 and 1935. FERA was supposed to fund relief programs in states with one dollar in federal aid for every three dollars in local assistance. The federal government ignored this formula in the plains, where the agency dispensed a total $145 million for general and special relief programs. Federal funds contributed 81 percent, local funds 18 percent, and states a scant 0.5 percent of this amount. At its height, in February 1935, 1 million people received general relief in the region.[27] Despite the obvious need for assistance, most plains newspapers and relief committees disdained direct relief. "Charity is repugnant to American tradition," claimed the *Lincoln (Nebr.) Star*, "the true citizen of this country desires to work, wants to stand on his own feet, insists upon performing hours of labor for the money he receives." The citizen "does not expect any one to give it to him. It is a humiliation to him, a blow to his pride, a debasement of his spirit. Always he has been independent, self-supporting, and able to take care of himself. Any other situation destroys his self respect."[28]

Though ideological considerations varied from state to state, county relief committees throughout the plains acted in the belief that direct relief hurt the esteem of its recipients in the long run more than it helped them. Families dependent on county relief were at the mercy of rigorous community standards. Across America, FERA doled out $25 to $29 to families to survive over a month. This was a paltry sum, since predepression industrial workers brought home this amount for *one week's* work. Direct relief was intended to provide only the barest of essentials. For example, local committees dispensing aid refused to pay for hospital care and rarely assisted indigent families with rent.

Direct relief was made more demeaning through its petty administrative procedures. The heads of families had to go through a means test to qualify. In addition, social workers visited clients' homes to investigate actual need and to evaluate how families were using assistance. Finally, much of the direct relief was given in kind, rather than money. This system was guaranteed to make urban and rural clients alike feel inferior and helpless for accepting aid from the government. When FERA ended in 1935, direct relief cases were thrown back to the limited resources and support of state and local governments.

The New Deal attempted to eradicate the need for direct relief through work relief programs. Supposedly, work relief clients would have more esteem and could remove themselves from the public dole by transferring to private industry when the economic recovery came. The first national attempt at work relief during the Roosevelt administration was the short-lived Civil Works Administration (CWA), which operated between November 1933 and March 1934. The CWA had no means test or demeaning investigation of its clients.

CWA jobs were supposed to pay as well as predepression jobs. But in 1934 the CWA paid a third less wages in the rural northern states than private employers.[29] The CWA in Nebraska, for example, allowed its clients to work only enough hours to cover their immediate needs, as set by the state relief committee.[30] Still, the CWA was successful at removing plains farmers from relief. From November 1933 through January 1934, nearly all Kingsbury County, South Dakota, "open country" families who left the direct relief rolls did so to work on the CWA.[31] Though the federally run and financed CWA provided much-needed wages for plains farmers, it was politically unviable for the United States. Just weeks after FERA administrator Harry Hopkins created the work relief program, President Roosevelt announced its discontinuation. The CWA was expensive to fund, and improved economic conditions during spring 1934 contributed to its demise.

The need for work relief remained strong, and within a year the federal government established the Works Progress Administration (WPA). Between May 1935 and July 1943 the WPA created millions of work relief jobs for Americans. Although New Dealers created the agency to provide work relief for the unemployed in towns and cities, it accepted farmers too during the times of extreme need. Throughout the plains, farmers worked for wages on rural projects such as grading roads and building runoff farm dams. Unlike the Civil Works Administration, the WPA did have a means test. And although the WPA was a federal program, its projects required local sponsorship. Over its eight-year life, local leaders criticized the WPA and Congress cut its budget. Conservatives saw it as a cesspool of political patronage and federal intrusion in local affairs. Harry Hopkins, who directed the program between 1935 and 1938, also had to face President Roosevelt's fiscal conservatism, periodic congressional cuts, and the country's fallacious expectations for economic recovery. Each made the WPA's existence uncertain.

Begun with the best intentions, the WPA sometimes demeaned its clients. Working for the WPA meant short hours, periodic layoffs, small monthly paychecks, and menial jobs. To benefit the most needy, federal regulations required 90 percent of all WPA workers to come from relief rolls. Periodic administrative clearings of relief rolls made financial stability for WPA workers impossible. Worst of all, the WPA provided jobs for only one-third of all those who needed work. The plains offered ample evidence of the WPA's shortcomings. As noted previously, many farmers in the region worked on other farms to supplement their income. During summer 1935 the WPA closed rural work relief cases in farm states such as the Dakotas and Kansas, partly because more substantial farmers feared that underemployed farm operators would rather work for the WPA than for them at harvest time. Even *Business Week* criticized this work suspension because families removed from the relief rolls could not resume their benefits after the harvest except by reapplying and submitting to new means investigations. Despite the abundant fields of grain that summer, however, few farmers could afford the extra labor.

A local USDA official predicted an ample labor supply.[32] A subsequent study in South Dakota agreed. Ninety-one percent of the surveyed 214 relief workers were normally farm operators. During the relief shutdown that summer, only 11 percent of the surveyed relief clients were able to find work in the harvest fields. The study also showed that these rural clients suffered a substantial drop in total income when they were bumped off relief. Before relief suspension, clients made a total monthly income of

$27.56 on average. Three-quarters of that amount came from relief aid. After suspension, the rural South Dakotans were left without an economic safety net, and their total monthly income fell nearly by half. These clients made up for the loss with funds from relatives, government food aid, homegrown food, and AAA payments. Clients turned the commodity checks over to creditors.[33]

Two years later, plains farmers who had left the WPA work crews had improved their situation slightly. One report surveyed North Dakota clients who were suspended or voluntarily left their WPA jobs between April and July 1937. The North Dakota rural clients appeared better off than families in other rural areas. Nationally, rural clients' monthly income dropped 20 percent, from $44 to $36, after they left the WPA. In North Dakota, former clients' monthly income stayed at $50 after leaving the work relief program. Like other rural Americans, 80 percent of ex–WPA workers in rural North Dakota had jobs by November 1937. Perhaps most surprising was that nearly all the surveyed American farmers formerly on relief remained farmers. Eighty-six percent of WPA workers who were farmers before they went on relief returned to the land after leaving the WPA.[34] Since the United States experienced a renewed economic recession between September 1937 and June 1938, there were few nonfarm jobs to pull men and women off the homestead.

The work relief programs saved many American families from poverty. Yet New Dealers purposely made direct relief less attractive than work relief, and work relief less attractive than private employment. The WPA program had two major defects for farm clients. First, despite the modicum of security afforded by work relief, the WPA system probably did far less for the morale of its clients than full-time skilled labor would have done. The WPA offered men and women on work relief inferior work, wages, and hours in labor-intensive jobs requiring little skill. Families could apply for work relief only after falling into destitution, so resorting to the WPA was a public admission of indigence. Second, workers' income from WPA wages was closer to direct relief than to paychecks from private employees. So work relief was closer to charity than employment. The New Deal job program was divided in its intentions. It tried to eradicate the stigma of relief but its substandard level of assistance pressured clients to find private work.

Borderline farmers were in a curious position within the relief programs. Farmers' unique problems and the nation's positive perception of them together demanded a program separate from direct aid or work relief. Although they were in need, farmers were not truly destitute, since they possessed property. But without their claim to their home, land, livestock, and equipment, farmers were truly paupers, materially and morally. Without the pride of ownership and independence in their livelihood, they were stripped of all they had worked for under normal conditions and had suffered to retain under the drought and economic depression of the 1930s. Plains farmers were particularly uncomfortable when accepting direct and work relief. They avoided the demeaning practices that came with relief when possible. In 1935, for example, FERA provided farmers of Ellis County, Kansas, with money to practice soil conservation. Yet in order to obtain the benefit, farmers had to swear they could not finance the plowing through their own resources. Farmers looked at this as an admission of destitution. One farmer commented, "Not 2 percent of the farmers of Ellis County will declare themselves paupers to obtain this federal aid."[35]

Plains social workers and farmers drew a distinction between aid for town and for the countryside. They, like most Americans, saw drought-stricken, impoverished farmers as the "deserving poor," who took aid only as a last resort. One Kansas caseworker commented that relief clients in town sought work relief. "The city man expects to be taken care of," wrote Julia Miller of the Kansas Emergency Relief Administration. Farmers, on the other hand, drove miles to find work. For a farmer to accept relief, "he must have gone his mile, I mean he must have mortgaged his land, [and] taken out all loans available, such as feed and crop loans."[36]

Many plains residents felt that aid for the borderline farmer was not an admission of sloth. Perhaps this was because borderline farmers, though often in debt and in need, were propertied entrepreneurs. The "undeserving poor" according to conventional wisdom, were shiftless, irresponsible, and lacked initiative. Borderline farmers, however, were men and women willing to work themselves out of debt and low income into middle-income status. Helping such a group of farmers was neither a waste of money nor a trap to make them dependent on the government. Rather, since what was good for the farmer was good for the country, it made good business sense to aid the men and women who drove the economy. As one child of a relief client said years later, "Well, we didn't consider it relief. We just considered it a subsidy. . . . we felt that the government was really on the ball to provide this sort of thing because it did help the farmers to get on their feet."[37]

Not surprisingly, the farm sector tried to temper the bitter taste of accepting assistance by rationalizing it. Whether people in the rural plains justified relief as an act of last resort, or as an economically sensible strategy, they tried to make their peace with it. This was difficult for men and women who saw themselves as independent entrepreneurs. The farmer and his representatives labored to come to terms with his new status as a relief client. One solution was to erase some of the humiliating paperwork. North Dakota's Senator Gerald Nye pressured the WPA to dispense drought aid to farmers without them having to register for relief as town clients did. "The requirement for registration has been most distasteful to many farmers who never have sought relief and do not seek it now," Nye said. Work would be available for farmers "without a tinge of 'relief' about it," the senator promised.[38]

Many farmers simply learned to accept relief in their own way. Grace Martin Highley, a relief director in western South Dakota during the 1930s, agreed that her neighbors found it hard to accept relief. However, Highley found that "men face up to reality and do what they can to survive. I think that people compromised. I think that you attach yourself to something that comes somewhere near your ideal, but in reality you reach out and take what you can get."[39] Still, it was difficult to admit that farmers or their neighbors had stooped to accepting relief, economic depression or not. One South Dakota schoolteacher recalled, "When I saw some of our respected, well-to-do farmers and businessmen manipulating a shovel on the WPA, it made [my] heart ache."[40] But for all its faults, work relief did keep many plains residents above water. During the 1936 presidential election the *Lincoln Star* insisted on keeping the WPA alive. The alternative to work relief, the Nebraska newspaper insisted, "would doom millions of employable men and women to continuous idleness, living on barely enough to keep soul and body together."[41]

The New Deal's direct relief and work relief programs implemented in the United States between 1933 and 1935 were unprecedented in both their scope and impact. The federal government had committed itself as never before to the welfare and employment of the American people. With this aid millions of families now were able to buy groceries, pay the rent, and generally avoid destitution. However, as operated in the plains, the limitations of direct relief and work relief were obvious. For borderline farmers especially, both forms of relief offered only a safety net against the worst financial hardships. Farm operators probably saw such relief as demeaning and poorly paying. If the WPA tried to engage the unemployed until industrial recovery, this strategy failed to benefit farmers—they were staying put in the countryside. What farmers needed was help that enabled them to stay *on* the farm, where they could provide for themselves, use their skills and resources, and ready themselves for the hoped-for recovery. Plains borderline farmers needed more than just to keep body and soul together. New Deal relief programs maintained farm families at the subsistence level but not much higher. However, these men and women on the margins of poverty wanted to return to their status as producers and increase their production, which required more than subsistence.

"Bold Persistent Experimentation"

By 1935 the Roosevelt administration realized that it had to do more than simply offer the American farmer relief, be it direct, work, or emergency relief. The American countryside, particularly the plains countryside, desperately needed "rehabilitation" to counteract drought, poor living and farming conditions, and debt. Drought was a relentless and pitiless visitor to the Great Plains throughout the 1930s. Conceivably, low commodity prices alone could not drive borderline farmers below the poverty line. Rather, persistently low levels of rainfall kept these farm families from realizing even a small degree of economic security. In the parched plains during the summer of 1935, drought drove 60 percent of all new relief cases to seek aid. These families in the Dakotas, Kansas, and Nebraska had to get by on the average monthly relief check of $19.[42]

During the Great Depression the Roosevelt administration came forward with a great variety of relief programs that assisted plains farmers. These ranged from the AAA program to other widely used programs such as grants and loans to farmers hurting from the devastation caused by drought and grasshoppers. The government offered to purchase cattle from farmers and ranchers with sun-scorched pastures and feed crops. This was essential for stockmen and women since delivering and selling cattle to conventional markets actually cost them more than the cattle were worth. Farmers could also borrow funds through emergency crop and feed loans to keep their operations afloat.

These purchases, grants, and loans allowed borderline farm families to survive. However, they did little to solve the longer-term problems of the maladjusted countryside. Plains farmers should have been able to pay off their government loans after a harvest or two. This was impossible during the mid-1930s. By the end of 1936, the federal government had made 369,000 emergency crop and feed loans, averaging $189 per loan in the Great Plains. Two-thirds of these loans were outstanding as of January 1937. In comparison, the average emergency farm loan in the United States was 42 percent smaller at $110, and was paid off at twice the rate. An amazing 58 percent of all emergency farm loans in the United States were made in the Great Plains.[43] The region was on its way to becoming the capital of government farm relief. The sympathetic policies of the Farm Credit Administration, which avoided foreclosing on farmers in arrears, alleviated the potential crisis of unpaid emergency farm loans. This was good news for plains counties such as Gray County, Kansas, where relief clients owed money to as many as eight separate government agencies. These were not transient "bums" on the dole. A survey of 182 relief cases in the western Kansas county found that nearly all clients were permanent residents who had never received relief before 1931.[44]

Faced with crop failure and low commodity prices, the only alternative for borderline farmers was to find a job on another farm or in town. Yet work was scarce. Because of the lack of jobs elsewhere, farmers requiring relief stayed home. A 1935 WPA study found that in the plains counties surveyed, 80 percent of all farmers on relief listed farming as their usual occupation. Only 18 percent of those who had been farmers were looking for work off the homestead. Nearly all client families had at least one gainful worker.[45]

Federal government farm and relief officials, especially those from the plains states, were faced with the seemingly intractable problem of a drought-stricken, immobile farm population sinking deeper and deeper in debt to private and government lenders. Clearly, a large sector of the plains countryside required an adjustment. Since the frontier days erratic environmental conditions and unstable income troubled the region. The land boom around World War I saddled many farmers in the region with debt.

During the Great Depression, the dependable option of leaving the farm for city jobs disappeared. Consequently, the plains borderline farmers demanded solutions other than those offered by urban relief programs. Their farms and homes were falling into disrepair and losing their productive capacity, yet on farms throughout the region there were men and women fit for work.

In response to the problems and opportunities of borderline farmers, New Dealers probed for different solutions. The extreme drought and economic depression of the 1930s required new answers—and new thinking. But the search for a consistent ideology supporting the new programs of the Roosevelt administration is truly a lesson in futility. President Roosevelt's New Deal programs reveal originality, pragmatism, political opportunism, and the expediency of working under emergency conditions. Roosevelt himself acknowledged the need for new solutions to meet the economic and environmental disasters that devastated America. Before the election, he proclaimed in Atlanta, "The country needs, and unless I mistake its temper, the country demands bold persistent experimentation. It is common sense to take a method and try it. If it fails, admit it frankly and try another. But above all, try something."[46] Kansas newspaper editor and Republican William Allen White accepted such experimentation with misgivings. The Sage of Emporia wrote Herbert Hoover in 1934 to offer his impressions of President Roosevelt. White observed, "The President gets on one running board after another headed for his evident desire to get out of the morass. . . . It is obviously a case of trial and error with the President . . . going a little piece down the road with anyone, backward and forward, zigzagging, covering and recovering." Despite this impetuous governing, White was convinced Roosevelt was "earnest and honest in his endeavor to get us out of the mess, though not intelligent."[47]

Out of the search for solutions for rural America arose the first coordinated federal government agency to address the long-term problems of the countryside at the grassroots level. The Resettlement Administration began in April 1935 and lasted until December 1936. From its inception it absorbed a bewildering number of projects and goals from other agencies. These included suburban community projects, migrant labor camps, and projects contending with stream pollution, seacoast erosion, and reforestation. The Resettlement Administration (RA) attempted to fundamentally reform and readjust the American countryside in several ways. First, as part of its visionary land reform, it attempted to retire vast expanses of American land ill-suited for long-term cultivation or grazing. Second, as part of this plan, the RA tried to resettle farm families from such marginal lands on government-built farms. Third, the agency sought to bolster and reform the farming and domestic practices of borderline farmers on their current farms.

Rexford Tugwell, the head of the RA, made an indelible impression on the rural rehabilitation program that survived well after he left the agency. Tugwell was born to a farm family in Sinclairville, New York. In 1931 he rose to a full professorship in economics at Columbia University. By the time of Franklin D. Roosevelt's election to the presidency, Tugwell had garnered a national reputation as an agricultural economist. In 1935 President Roosevelt appointed him head of the newly formed Resettlement Administration. Tugwell's personality and philosophy distinguished him from other New Dealers. Tugwell was politically well to the left of other members of Roosevelt's Brain Trust, not to mention most Americans. Tugwell maintained that the federal government should be more than a regulator of private enterprise. He believed it was the only institution that could feasibly determine and execute national economic plans and manage natural resources. The professor of economics was convinced that uncontrolled private enterprise in American agriculture crippled the country-side. What set Tugwell apart from other New Dealers was his willingness, particularly in the farm sector, to use the federal government as a vehicle for social and economic reform. His opposition to the competitive impulses that animated American commerce swayed the development of the Resettlement Administration. It also made that agency and its successor, the Farm Security Administration, easy targets for critics who were already suspicious of government involvement in the countryside.

Other New Dealers had been at work on the problem of the destitute countryside. They saw the need for a separate relief program for insolvent farmers, since FERA was oriented to the problems of urban-industrial America. Under President Roosevelt, who often assigned one task to several government bodies, the New Deal set up at least three different agencies to treat rural poverty. These were the Department of the Interior's Subsistence Homestead Division, FERA's Land Program, and the AAA's

Land Policy Section. Each of these agencies attempted to specifically address the problems of rural poverty. In April 1934 the federal government ruled that rural rehabilitation programs would replace work and direct relief in rural areas.

During a trip through the Midwest in 1935, Tugwell conceived of a large government agency to unite the scattered government programs working to remake rural America. One of his goals was to create a new agricultural extension service to aid farmers on the lower margins of rural society. Acting on Tugwell's concept later that year, President Roosevelt created the Resettlement Administration through executive order. It was initially funded by the Emergency Relief Appropriation Act of 1935, which up to that time was the largest single appropriation ever made by the federal government. The Resettlement Administration from its birth faced two administrative problems. First, it was saddled with a myriad of former federal agencies set up to adjust marginal farm land and marginal farm families. Second, the WPA tried to transform the Resettlement Administration into a rural relief agency. Harry Hopkins attempted to bequeath counties with towns no larger than twenty-five hundred people to Tugwell, who rejected this plan. He planned to remake the countryside, not run a general relief agency for rural America.

The Resettlement Administration sprang not only from the agricultural and general economic conditions of the 1930s but from USDA rural social science experts during the 1920s and 1930s who helped shape it. Many of these experts had seen hundreds of thousands of ill-equipped farm families living on poorly used and infertile soil in the American countryside. After World War I a corps of "service intellectuals" arose within the USDA to deal with the long-term problems that troubled the plains countryside. Henry C. Taylor, the influential head of the Department of Agriculture's Bureau of Agricultural Economics (BAE), and other agricultural economists observed that many plains lower-income farmers seemed incapable of adjusting their operations to the changing needs of agriculture. Taylor determined that there existed a "degenerative" class of such farmers. He feared that the American countryside would be the dumping ground of such men and women.

Taylor greatly influenced Milburn L. Wilson, head of the Division of Subsistence Homesteads and later undersecretary of the USDA. In 1933 the Subsistence Homestead Division began pilot projects where the federal government provided land, farm implements, housing, and machinery to the urban and rural unemployed to start their own farms. Wilson had gained special experience in rehabilitating plains farms when he ran the Fairway Farms project in Montana during the 1920s. In this project Wilson attempted to prove tenant farmers who had lost their land after World War I could run successful family farms in the semiarid Great Plains. The project bought abandoned land, set up terms to lease the land under a tenant-purchase contract, and provided farmers with appropriate equipment. These operations were not small; they ranged from 640 to 2,640 acres of land, in line with other farms and ranches in the region.[48] Fairway Farms was clearly designed for the propertied borderline farmers.

Through his experience, Milburn Wilson was convinced that "progressive" farmers could run successful operations complemented by modern conveniences in their homes. He described progressive farmers as those who kept up with managerial, scientific, and technological advances by subscribing to farm magazines and attending farm extension courses. In contrast, Wilson and other members of the Bureau of Agricultural Economics also identified a large group of poor, unprogressive farmers. Through their lack of resources and skills, these submarginal farmers, the planners believed, had little chance for success.

BAE economists deemed that too many of these peripheral farms dotted the countryside. The agency determined the poorer farms by factors such as climate, soil quality, farm size, capital costs, and, importantly, the ability to increase the scale of farm operations. Milburn Wilson and his fellow BAE economists believed that aiding submarginal farmers on submarginal lands harmed the American countryside. Accordingly, many USDA insiders believed that certain soil-depleted areas should be taken out of cultivation. Small farms with few resources and little chance to grow, they argued, should be phased out by government farm programs. In response to these problems, the USDA's Land Planning Committee, after completing an inventory of the nation's resources, recommended a huge reduction in the nation's cultivated land and rural population. Under their plan, the federal government would procure 75 million acres of "submarginal" farm land (7 percent of the country's farm land) and remove it from full production. Such an operation would potentially displace 450,000 farm families.[49]

Government planners took notice of the difficult farming conditions in the plains, too. The federal government reported that the lack of rain and soil exhaustion demanded "certain readjustments between the people and the land." More than 36,000 families had already moved out of the Great Plains drought area. An estimated 59,000 families remaining in the region made up the "surplus population" requiring relocation, according to these planners. Therefore, rather than attempt to aid all income levels of the countryside, many USDA officials had already written off large numbers of farm families who had fewer resources and less land than borderline farmers.

The plains appeared to have two classes of troubled farmers: those who were leaving or trying to leave the region, and those who remained. For the mobile class of families, the government offered nothing. With all the obstacles to farm life during the Great Depression, leaving was understandable. It would have made little sense to encourage the departed to return, or to maintain other families on burnt-out, impoverished farms once they made the conscious choice to move. The second group of troubled farmers was the "stickers." For financial, social, and personal reasons they stayed, despite poor farming conditions. Within this battered, immobile class, the USDA found both "surplus" and "progressive" farmers. Based on the evidence of the time, it was reasonable to conclude that the plains could not support all these farmers. Though it was perhaps callous, it made sense to divide the remaining men and women into those with the skills and resources to run a profitable operation, and those who did not. This kind of "triage" was necessary in a region where the land had consistently failed to sustain many rural families at an acceptable standard of living. The Roosevelt administration created the RA on behalf of borderline farmers who could benefit from such aid.

There were two conceivable strategies to helping these plains borderline farmers. First, the federal government could move them off their land to a better operation. This was the thinking behind the RA's land retirement and resettlement programs. Second, the government could assist these farmers where they were, using the rural rehabilitation program. In the mid-1930s, the Great Plains was a disaster area. Drought, crop failure, insolvency, and threats of mass destitution marked the plains as a problem region. If the nation needed a reminder of this, a May 1934 dust storm blew an estimated 300 million tons of soil from the plains, causing around $5 billion in damages. Congressman Clifford Hope, who represented the western Kansas district pummeled by drought, wrote that on May 12 the storms "showered dust on the Capitol and White House. Some was said to have settled on the President's desk. When I left the Capitol that evening, I found my car covered with a film of familiar-looking Kansas soil."[50]

Dust from the plains had followed President Roosevelt to his office in 1934, and he sought to help the area. That year, the primary sources of income for drought-stricken farmers in the region were government cattle purchases, AAA payments, work relief, and direct relief. President Roosevelt took a personal interest in plains farmers and considered their chief problem to be cultivation of crops in a marginal, semiarid region. The New Deal's solution to this problem was to subsidize soil-conserving farm practices, tree planting, and the restoration of grazing lands. Furthermore, USDA planners suggested taking certain areas out of cultivation altogether and removing local farmers to more suitable lands. Significantly, Congress never acted on the latter readjustment plan.

At the national level, planners were open to halting agricultural expansion into arid and semiarid areas. After the land boom following World War I, farm experts criticized wheat growers' substantial expansion into semiarid lands. However, this criticism arose from concern about the surplus of grains during the twenties rather than from environmental concerns. By the 1932 presidential election, both national parties urged some sort of regulation of agricultural land use. Eventually, under the various New Deal agencies, the government purchased 11 million acres of land in the country. Of the twenty-four thousand families living on this land, only 9 percent were resettled with government assistance.[51] The rest apparently remained on the property in limbo or moved away without government assistance.

During the mid-1930s land retirement had considerable support in the semiarid plains. At the local level, town and county governments saw the positive side of the government purchase and retirement of parched lands. Land sales by farmers to the government were voluntary, and under the program the government compensated counties for lost property taxes with receipts from grazing fees on retired lands. County officials hoped that this readjustment would lower their relief expenses. Faced with thousands of families leaving the region, this was the best alternative around.

Particularly between 1935 and 1937 local press and individuals supported land retirement. In summer 1935 the *Rapid City (South Dakota) Journal* noted that local landowners had acquired options from the government to buy nearly nine hundred tracts of land totaling ninety thousand acres. This was, according to an editorial, "definite proof that a large number of individuals owning land in the area want to sell it." However, the newspaper closed, "whether or not the purchase of the land by the government is proper is a debatable question."[52] Absentee landowners were especially eager to sell out to the government. In May 1935 E. R. Jonson of Illinois wrote the Resettlement Administration that he had been unable to pay taxes or the mortgage interest on land in Kidder County, North Dakota. Jonson declared, "In my desperation, I am wondering if there is any [land retirement] project being planned for that portion of N. Dakota. . . . Anxiously awaiting a reply from you, and assuring you I will be truly grateful for any word, no matter how meager."[53] One woman wrote to sell a 640-acre farm in the same North Dakota county so that she and her husband could purchase a dairy farm in Vermont. The land, which she received through an estate, still had nearly $400 in back taxes on it. "I haven't had enough to live on let alone pay taxes," she wrote. "Will you be so good as to examine this property and if it is at all possible purchase it for reclamation[?] I will accept almost any amount. $5.00 an acre would be a Godsend." She closed saying that although the land had not produced crops for five years, it was mortgaged at one time for $10,500.[54]

By the middle of 1936 the Resettlement Administration had accepted options from landowners to purchase nearly 2 million acres in the Dakotas and Nebraska. Of that land, 470,000 acres had been purchased. The largest area optioned at the time, with nearly 700,000 acres, was in McKenzie and Golden Valley Counties, in western North Dakota.[55] Reactions from this area give a sense of both the support and eventual opposition to land retirement in the plains. In 1935 southwestern North Dakota had high relief needs, tax delinquency, and low crop yields. Counties such as these in the semiarid plains seemed to fit the land retirement plan perfectly. One local RA official boasted that because of federal optioning of local land, "we will now have an excellent chance of demonstrating the most productive use of land in the area."[56] Unfortunately, the land retirement program left families uncertain of their status on the optioned property. After eight months only 6 percent of the optioned land had been purchased—leaving farmers on the property unsure of their future. Another problem was the impact of the federal land purchases on local businesses. In late 1938, Rhame, North Dakota, made national headlines when its businesses demanded the federal government purchase the entire town. Because of nearby land retirement projects, the Rhame Commercial Club protested that "practically all the businessmen are facing financial disaster." Future land purchases meant shrinking property values and greater migration from the area, which would, according to the club, render local business and residential property "practically valueless."[57]

What was to become of farm families on retired land? Various New Deal programs purchased land and erected farm homes and buildings throughout the plains on which to resettle these landless families. The RA eventually took over these "rehabilitation homesteads." In Nebraska, Governor Robert Cochran visited a homestead project near Kearney in fall 1935. After inspection, the governor praised it as "modern pioneering." A project client, Joe Suchy, a forty-eight-year-old father of five, showed his appreciation, too. In 1934, he said, "we were broke. I had lost my farm, we had moved into town and I couldn't find steady work. Things looked awfully dark." Now, with his small truck farm he had money and a comfortable house. "Next year we'll do even better for we are learning as we go. The chance has meant everything to us."[58]

However, in Nebraska, as in the rest of the plains, the rehabilitation homestead program was controversial for two reasons. First, these were expensive model farm homes. The average plains homestead farm, with buildings, home, and land, cost $10,000.[59] Second, the farms' size was sometimes inappropriate to the needs of plains agriculture. For example, when they began, Nebraska's project farmsteads were a minuscule eighteen acres in size.[60] Even for truck farming, this was insufficient. Plains agriculture meant spreading small investments over large acreages of grain. For dryland farmers, the project irrigation plots required skills and experiences they lacked. In summer 1935 the *Omaha World-Herald* published a letter from a Douglas County, Nebraska, farmer criticizing a newly built local homestead project. Henry Glissman asserted that the project's fields were poor, small, useless for irrigation, and mosquito-ridden. Client farmers could not make a living on the small parcels of the expensive homesteads, Glissman wrote. Then the farmer predicted "these homes will be abandoned and stand as a gruesome monument to a government's inefficiency and folly in fostering a movement that, to a practical mind, had the earmarks of a failure from the start."[61] Four months later, the project, which the newspaper disparaged as "FERAville," was still not completed.

The *World-Herald* claimed that the project, under "generalissimo" Rexford Tugwell, was in a "state of coma."[62]

Though Mr. Suchy's and Mr. Glissman's statements may have been coached, they do reveal that homestead clients probably were grateful for the opportunity to relocate on new and stable property, while others were suspicious of government programs. Because of such mixed reactions from the public, by 1936 the plains land retirement and farm resettlement projects were stillborn. Despite the urgency of the region's problems, the sponsored migration of tens of thousands of farm families was too much for the Roosevelt administration to propose for both financial and political reasons. Therefore, the Great Plains Committee, which President Roosevelt had commissioned to investigate the region and its problems, made only moderate recommendations. These included keeping unbroken lands from cultivation, purchasing "submarginal" farms, then resettling its farmers and converting the land to pasturage, and establishing county committees to implement soil conservation practices.

RA chief Rexford Tugwell at first supported plans for a mammoth land retirement and farm resettlement program. However, because of public pressure he endorsed the Great Plains Committee's recommendations, countering that his agency would not finance farmers' "indiscriminate moving" to a different region. Rather, Tugwell offered the conservative solution of loans averaging $350 to repair their farm buildings and purchase seed and machinery. The loan would be secured by a crop or mortgage on their possessions. Tugwell also emphasized that the RA would act as an information agency to prevent an "exodus" of farmers.[63] The expense of resettling the thousands of plains farm families living in drought areas, the return of meager rainfall in 1937, and the natural inclination of farm families to stay put undercut plans for mass "resettlement." The director of the National Emergency Council summarized the situation to President Roosevelt in the scorching month of August 1936. The Dakotas in particular had widespread crop losses and shortages of water for livestock. The report recommended feed, seed, and subsistence loans to keep farms operating and the livestock alive—but no abandonment of the plains.[64] Therefore, Tugwell shifted the agency's land retirement and resettlement plans to rural rehabilitation. The Resettlement Administration's mission, he wrote, was to fight the "arresting forces which are causing erosion—human as well as soil—and of deterring trends which are depleting both human and natural resources."[65]

Even after they abandoned large-scale land retirement and farm resettlement schemes, Rexford Tugwell and other agricultural economists wished to balance America's land resources with the needs of millions of farm families living on the margins of poverty and environmental instability. Market forces and maladjusted farming practices in semiarid areas forced thousands of farm families off the farm or into insolvency by the time the Resettlement Administration was created. Now that market capitalism floundered in the region, Tugwell and others argued, government planning deserved a chance.

However, Tugwell and the RA confronted a fundamental problem executing these goals because of the way the agency was created and funded. This sort of a government agency, with its great potential for transforming rural America, should have passed before the U.S. Congress for individual consideration. The agency would have benefited from public hearings on its goals, administration, and funding, allowing the RA to assemble interest groups and a constituency for future support and advice. While New Dealers conceived the AAA, for example, they also garnered the approval of important farm organizations and legislators from rural states for the agency. In contrast, the Resettlement Administration arose from a presidential executive order, based on a vague congressional mandate. Unlike other agricultural agencies, the RA and FSA were subsequently funded through emergency appropriation bills until 1942. Under the emergency situation, it was understandable that agricultural experts formed rural rehabilitation in haste without public review. However, to many farm observers the rural rehabilitation program was formulated in secret and run on the sly. This contributed to the RA's, and later the FSA's reputations as nebulous emergency relief agencies conjured by a Brain Truster and lacking widespread support among the press, farm leaders, organizations, and legislators from farm states.

To those inside the government, the goals and organization of the Resettlement Administration were often unclear. Roosevelt's secretary of the treasury, Henry Morgenthau Jr., recalled a meeting with Harry Hopkins, the relief czar, on funding for the RA in April 1936. President Roosevelt had apparently requested that the Resettlement Administration and the

WPA be funded through the same $1.5 billion allocation. Hopkins, who was supposed to testify in favor of the bill, was confused about how to describe the new agency to a House committee. "I do not want to say anything that will be unfair to Resettlement," said Hopkins, "I cannot go up there and say that I do not know what the President has in mind."[66] Others within the USDA were convinced that the Resettlement Administration was Tugwell's castle in the sand. Howard Tolley, who later headed the Bureau of Agricultural Economics, claimed that Tugwell had offered him the second spot in the Resettlement Administration. Tolley turned it down. He felt the agency "was going to be another wild scramble poorly organized, difficult to administer and difficult to get anything done in."[67]

Between 1930 and 1936, while drought and economic depression burdened the plains, state and federal governments and farmers fashioned a new relationship with each other. Herbert Hoover proved unwilling to allow much aid beyond supporting large marketing and credit institutions. Initially, the New Dealers under President Roosevelt sought to control the rise in mortgage foreclosures and to resuscitate the buying power of the American farmer. Yet the benefits of farm programs of the early New Deal failed to put borderline farmers on stable ground. While AAA checks helped them pay late bills and back taxes, they remained insolvent. In the meantime, millions of federal dollars poured into the plains for feed and seed loans, cattle purchases, and subsistence grants that kept borderline farm families from outright destitution. Yet they still needed to adjust to the long-term problem of the increasing cost and scale of plains farming. New Dealers sought ways to sustain the indigent through direct relief and work relief. While relief programs kept farm families from outright privation, assistance did nothing to maintain farm productivity once economic recovery occurred. Through the interplay of conservative and liberal notions of state and society, the financial constraints of government, and the hard realities of the thirties, the rural rehabilitation program was born.

Though the relationship between farmers and the federal government changed with the New Deal, no one wanted to change the rules of rural capitalism. Leftist New Dealers such as Rexford Tugwell wanted to reform business practices and resource use and allocation. Tugwell wanted sweeping national planning for rural America. He would have approved of a sep-

arate extension service for borderline farmers emerging from rural rehabilitation to counter the power of the conservative Farm Bureau. Moderate
New Dealers, such as USDA Secretary Henry A. Wallace, wanted a financial
safety net for middle- and upper-income farmers until prosperity returned.
Few within the USDA expected government commodity payments to last
beyond the economic depression. New Dealers and the general public saw
alphabet programs such as the WPA, the RA, and later the FSA as emergency
measures to maintain and improve families during the disastrous early
1930s. This explains why these agencies were dropped by a Democratically
controlled Congress during World War II. The next chapter explores rural
rehabilitation at work in the Great Plains.

4

"Rehabbers" in the Great Plains

Helping the "Mendicant Plains"

A Kansas farm woman, Mrs. W. L. Hannon, sent a letter to Eleanor Roosevelt in 1939. At the beginning of the decade, the woman wrote, her prosperous farm family had "youth, health, and ambition." Their success during rough economic times encouraged them to float loans to purchase cattle. Then, she wrote, "we awoke one morning to find we owed money we never had possessed." This was only the beginning of their problems. At first the couple expected an economic recovery. Unfortunately, after years of drought, weighted down by $3,000 in debt and harassed by unsympathetic bank collectors, Mr. Hannon's health broke under fatigue and anxiety. Then the couple learned of an innovative New Deal program and they found salvation in a loan from the Farm Security Administration. With the loan the Hannons had "a new lease on life." In addition, the bookkeeping required by the loans helped make "a businessman of the farmer." Mrs. Hannon sincerely appreciated the creative program of grants and loans to place farmers on their feet again. She closed the letter to the First Lady by writing, "Not a day of my life I do not thank God for President Roosevelt and his leaders. . . . From a grateful farmer and his wife to whom the Farm Security Administration has given health, happiness, and courage."[1]

Thus far we have explored the circumstances that made the Great Plains an ill-fated agricultural region in America. Economic and environmental forces during the 1930s especially harmed borderline farmers. They were entrepreneurs who lacked the income, tractor power, and additional credit to dig themselves out of debt. Especially hard-pressed were farm tenants, who lacked agricultural, economic, and social stability. After nearly six years of financial hardship during the Hoover and Roosevelt presidencies, the Resettlement Administration, then the Farm Security Administration, arose to help these farmers achieve financial stability through the rural rehabilitation program. Unlike the Farm Credit Administration and the Home Owners' Loan Corporation, which helped farmers refinance their farms and home mortgagees finance their homes, the RA and FSA offered subsistence and operating credit for farmers.

Mrs. Hannon's heartfelt testimonial celebrated the positive influence of an activist New Deal agency. Between 1935 and 1944 the rural rehabilitation program gave millions of American farm families a "new lease on life." This is especially true in the Great Plains. Beginning in 1935 and leading up through the war years, ninety-five thousand plains families received special loans, and seventy-three thousand received cash grants from the two agencies.[2]

Historians have examined both the general New Deal farm programs and the rural rehabilitation programs. Anthony Badger looks at the New Deal in rural America as "a holding operation for a large underemployed agricultural labor force." For more substantial farmers, the New Deal eliminated some of the risks in agriculture. Through programs such as the Agricultural Adjustment Administration, which paid farmers for taking land out of production, larger farmers could "organize effectively to take advantage of government intervention." The real goal during the Roosevelt years, according to Badger, was to correct farming through planning and efficiency. Large-scale farmers still in operation during World War II eventually benefited from bountiful rains and improved commodity markets.[3] Historian David Danbom emphasizes that larger farmers specializing in a few commodities profited more than smaller, more diversified farms from New Deal agricultural programs. He contends that out of guilt for leaving the borderline farmers behind, Roosevelt financed the RA and the FSA with "conscience money." These agencies, in Danbom's words, performed "a sort of agricultural triage, saving a handful of relatively promising farmers and letting the rest go."[4]

Indeed, because of the lack of political support and funding, the rural rehabilitation program ignored thousands of lower-income plains farm families in the 1930s. This group consisted of a mixed lot of part-time farmers, struggling tenants, retired and young couples, and the poor. They had too few resources in land, livestock, machinery, or capital to be "rehabilitable" by the government's definition. These people had to rely on their own means and private and general government relief programs to survive.

Despite the shortcomings of the rural rehabilitation program, it was reasonable and necessary in its intent and operations. Never before had an institution, private or public, attempted to actively improve farming conditions and the rural standard of living. With hindsight, it is easy to dismiss programs that attempted to maintain or improve by small measures the lives of borderline families. Unlike the great public works projects of the era, there are no tangible monuments to the subsistence or incremental increase in the livelihood of "promising farmers." Success in the rural rehabilitation program came in the form of a few more dollars at the end of the year. This money went into retiring debt, purchasing a few consumer goods, or reinvesting in the farm. Or in not having to pack up and move to another tenant farm, as Mrs. Hannon of El Dorado, Kansas, could attest.

Before the plains and its borderline farmers were rehabilitated, the region had to come to terms with outside assistance. Despite its independent, self-reliant image, the American West during the Great Depression avidly sought money from New Deal programs. Such demands were not new, as the region had long been dependent on federal largesse. As writer Bernard DeVoto noted in 1934, the West had been "mendicant and rapacious" in its clamor for aid since first settled by whites. DeVoto added that the region had a special penchant for extending its dreams of wealth beyond its resources. This expansionary drive often left the West in the role of needy stepchild to the wealthy East during hard times. "It is the section of the country," DeVoto wrote, "in which bankruptcy, both actuarial and absolute, has been the determining condition from the start."[5]

This was true, since the West proved the most successful region at attracting New Deal dollars, particularly federal farm aid. For example, Alf Landon, Republican governor of Kansas, effectively managed to steer funds from New Deal programs to the his state. Opponents even accused Landon of balancing the state's budget by shifting the burden for relief to Washington DC. Clearly, the West was willing and able to tap into the wealth of the national government. Influential Republican legislators such as Arthur Capper and Clifford Hope of Kansas, George Norris of Nebraska, Lynn Frazier and Gerald Nye of North Dakota, and Peter Norbeck of South Dakota deftly diverted New Deal money to the Great Plains.

The West was not America's poorest region during the 1930s; that dubious honor went to the South. However, the West was the neediest area in terms of lost income. There was a correlation between the drop in per capita personal income during the Hoover years and high unemployment, and the per capita New Deal aid received in the western states. The plains were particularly hard hit by plummeting income during the thirties. North and South Dakota, for example, had the first and second most profound drops in per capita personal income in the nation during the Great Depression. Plains per capita personal income fell 52 percent, from $445 to $231, during Hoover's administration. Therefore, the plains states, because they lost the most income and because of political pressure from the region, netted the most in federal aid during the 1930s. While the nation's per capita federal expenditures amounted to $224 between 1933 and 1939, the Great Plains averaged $363, or 38 percent more, than the nation as a whole.[6] Rural rehabilitation clients shared in their region's good fortune. For example, in 1940 the average plains client had $600 in credit from the FSA. In contrast, the average Mississippi client family had only $330 in credit that year.[7] Racial prejudice toward African Americans among local rural rehabilitation committees probably caused this.

Rocky Beginnings for Rural Rehabilitation

The Resettlement Administration and the Farm Security Administration were, like other New Deal programs, haphazardly formed and run during their early years. New Dealers themselves admitted this. Secretary of Agriculture Henry A. Wallace confided to his diary in 1935 that in the Roosevelt administration "the objectives are experimental and not clearly stated. [There are] many advantages to this approach but it does not lead to the happiest personal relationships and the best administration."[8] The two agencies were agglomerations of earlier rural relief and land adjustment programs. As a result, both seemed to have several goals but little direction. As Will Alexander, who became head of the FSA in 1937 noted, the agency was "a mess. They'd just been dumping everything in it. Administratively it was an impossible kind of thing for that time." Characteristically, Alexander learned he was the new head of the FSA through the press; President Roosevelt had not discussed the position with him.[9] So, despite the careful study and planning that went into creating the RA and its policies, the lack of initial funding and the poor coordination with other agencies hurt its debut in the plains.

To be fair, the early Resettlement Administration had an extraordinarily difficult task. It was attempting to fuse together the disparate New Deal initiatives meant to provide relief and to reclaim an American countryside pummeled by five years of low agricultural prices and merciless drought. The Resettlement Administration's rapid growth exacerbated its organizational problems. When the agency began in spring 1935, it had a staff of twelve; by the end of that year it had more than sixteen thousand employees. Along with the proliferation of workers came the propagation of many rules and regulations. This novel government agency had a difficult time keeping up with its own activities. Not only was the RA involved in the rehabilitation of borderline farmers, but it also implemented and managed various land utilization and farm resettlement programs. Administrators had difficulty keeping track of their divisions, which became increasingly elaborate. As a result, field supervisors often lacked a sense of direction in their duties.

Charged with resuscitating insolvent farm families, the RA became a hothouse of innovative ideas. The Resettlement Administration had a reputation as something of a liberal think tank for rural America. In fact, many in the USDA regarded the agency as a "thinking" rather than a "doing" department. This had attendant benefits and drawbacks. On one hand, the RA and its successor, the FSA, were surprisingly adaptive to the changing and confusing conditions of rural America during the late 1930s and early 1940s. On the other hand, the agency was criticized as overly innovative for a traditionally conservative sector of America. Will Alexander admitted that his enthusiasm for starting programs eclipsed his managerial abilities. "I never wanted to administer anything," he said. "I liked to get ideas and get them rooted somewhere, but as far as sprinkling the plants for buds after the plants start to grow—that sort of thing [didn't] interest me very much."[10]

Despite urgent demands, the new RA lagged behind the needs of plains farmers. Although the Resettlement Administration officially began in April 1935, many of its operations were not in place in the plains states until September. Even then, poor coordination among federal agencies caused many struggling farm families to fall through the cracks. For example, local officials dropped farmers from federal work and direct relief programs and sent them to the local Resettlement Administration offices before they were functioning. Federal direct relief ended in December 1935, when the

Federal Emergency Relief Administration ceased operations. This threw the responsibility for caring for indigent farmers back to the states, counties, and communities. The result was confusion as state officials scrambled throughout the late summer and early fall to meet the needs of farm families. In August 1935 the South Dakota welfare office claimed they had neither the authority nor the funds to provide aid for rural clients. After a plea from Governor Tom Berry, RA head Rexford Tugwell released aid so that South Dakota farmers received $468,000 in emergency loans, averaging $16 a family, apparently for subsistence needs.[11]

Kansas farmers also faced destitution without aid from the Resettlement Administration in September 1935. Under initial government guidelines, farmers could not receive further government loans without repaying past ones. This was clearly impossible for insolvent farmers. The *Topeka State Journal* objected that "Between the regulations of the rehabilitation corporation and the seed wheat corporation, there is a vacuum spot where relief agents don't seem to be able to do anything but worry. . . . Somewhere between Mr. Wallace and Mr. Tugwell and Mr. Hopkins and Mr. Ickes . . . there is a confusion of requirements and red tape." The newspaper complained, "In the meantime . . . 17,000 farm families don't know where their winter groceries will come from and they don't know how to get feed for livestock or find money for taxes."[12]

That month Kansas congressman Frank Carlson claimed the federal government had forgotten the farmers. Characteristic of western representatives, Carlson danced an intricate shuffle, demanding aid while shirking responsibility for it. While asserting that "Kansas farmers do not want a dole," he called for work relief through the WPA and money for farmers to pay their feed and seed bills.[13] At the end of the month, Representative Clifford Hope of western Kansas also demanded more assistance. Hope complained about the "critical situation" in his district among rural rehabilitation clients unable to get advances to purchase wheat seed for planting next year's crops. Crop failures in 1935 had forced many farmers who had managed to stay off relief to apply to the rural rehabilitation program. No aid was available, however, until November, when it would be too late for planting.[14]

North Dakota in late 1935 and early 1936 provides a case study of the rural rehabilitation program's difficulties. According to the state's director of the Resettlement Administration, Howard Wood, the agency failed until

December 1935 to set up the proper facilities. When FERA ceased its direct relief activities in late 1935, farmers in the western counties were certified for WPA work—only to be told that they were ineligible. They were then referred to the state's fledgling Resettlement Administration. Director Wood cited other problems, including the lack of clerical assistance and harsh winter weather. A special obstacle lay in relief supervision at the county level. With the demise of FERA, county relief staffers were laid off. In addition, rural rehabilitation supervisors found setting up satisfactory farm plans quite difficult for farmers who had amassed sizable debts.[15]

Critics of the early rural rehabilitation program complained about its expense. Administrative costs made up 15 percent of its national budget and 8 percent of the total budget for the Great Plains' loan and grant programs during the late 1930s.[16] This was high for a government body, yet not exorbitant given the cost of rehabilitating individual farm families. Yet there were definite signs that rural rehabilitation lacked the funds to cover rural insolvency in states like Nebraska. In April 1936 the secretary for the Oxford, Nebraska, Chamber of Commerce, H. M. Pettygrove, wrote to President Roosevelt about the "critical" circumstances of rural rehabilitation in his state. Under current funding, Pettygrove wrote, only one in twelve of the state's applicants received assistance. He cited John Whipple of Arapahoe, Nebraska, who ran a 480-acre operation with a total debt of $4,300. During flooding in the Republican River valley, Whipple lost nearly all his farm equipment, work stock, and most of his cattle herd worth $10,000. Pettygrove considered Whipple "frugal, a good business man, honest, dependable and entirely worthy in every respect." Commercial banks usually denied credit to farmers in such circumstances, referring to such loans as "charity." Unfortunately, because of a shortage of funds the farmer could not receive federal aid. Pettygrove pleaded, "The nation that fails to take care of such cases as his, isn't worthy of the name of a nation."[17] One month later, the state director of rural rehabilitation, L. A. White, wrote Nebraska governor Robert Cochran to explain the situation. The state of Nebraska initially received only half of its expected rural rehabilitation appropriation. "Lack of additional funds," White wrote, disappointed many clients who "naturally [felt] that their cases are just as urgent and worthy as those whose loans were completed."[18]

Finding the right staff and developing viable goals was critical to the plains rural rehabilitation program. However, creating and running such a novel program in the plains was a demanding task. Under the best of circumstances, it would have been difficult to find a manager to head the multifaceted Resettlement Administration/Farm Security Administration in the plains. The head was responsible for the intransigent farm poverty in the volatile plains environment during its worst drought and economic depression ever. In fall 1935 Rexford Tugwell picked Cal Ward of Douglas County, Kansas, to head the Resettlement Administration's Region VII, which covered the Dakotas, Nebraska, and Kansas by the early 1940s. Ward, who served as Region VII's head from 1935 to 1942, had been president of the Kansas Farmers Union since 1929; he then served as a regional advisor to the AAA. Rexford Tugwell called Ward a "progressive Republican." Despite Ward's party status, Tugwell assured President Roosevelt that he was "entirely in sympathy with the aims and purposes of this Administration."[19] The *Kansas Union Farmer* credited Ward's appointment to "his demonstrated ability to understand and analyze the problems of the common dirt farmers, and because of his record as a conservative progressive in matters affecting the economic welfare of farmers." The *Union Farmer* noted that Ward had worked with members of other farm organizations and nonfarm representatives to fight for the "economic and social equality of agriculture with other industries."[20]

Even with its machinery in place, the rural rehabilitation program was a mixture of ideological inconsistency, changing and illogical goals, and varying interpretations of its mission. Naming a "conservative progressive" to lead the plains rural rehabilitation program met with the region's consensus. However, the agency's staff at the national, state, and local levels often disagreed with one another. The ranks of national rural rehabilitation planners, state bureaucrats, and local supervisors revealed the schism between its goals and its staff. At the national level, the RA's idealistic planners were former AAA staffers who left that agency when they felt it favored larger farmers. Many of them were liberal New Dealers. The upper leadership of the plains rural rehabilitation program consisted of educators, businessmen, and attorneys.[21] The farm and home supervisors were often graduates of agriculture schools, former bank officers, and women with home economics degrees. Few had education or experience in social work.

Two important considerations concerning state- and county-level plains rural rehabilitation staffers hampered the program's reputation and operations. First, the federal government treated RA and FSA county supervisors as second-class government bureaucrats. They were denied civil service status and paid less than their counterparts in the AAA and the Soil Conservation Service. Second, many of the farm and home rehabilitation supervisors were Republicans. In the conservative, normally Republican plains, it was difficult to find local staffers totally committed to the New Deal. This was a source of unending frustration for county Democratic chairmen throughout the region. During the 1940 presidential election, for instance, the chairman for the Osborne County, Kansas, Democrats wrote the Democratic National Committee that not all the opposition to the New Deal in his county came from outsiders. Rather, "in the agencies set up by the government . . . many of the employees were quietly active against the Democratic Party." This bred antagonism with the Democrats' national rural rehabilitation program. In the local FSA office, rather than trying to explain administrative problems and calm the clients, the chairman reported, FSA staffers "would simply shrug their shoulders and say 'Government regulations, blame the government, not us.'"[22] That year, the Democratic chairman of Beadle County, South Dakota, bitterly criticized Republicans in the New Deal farm program in the plains. The GOP staffers infiltrated the agencies "like a bunch of termites [and] they ate out the foundation and devoured most of the superstructure. Through lies and deception they undermined the good opinion that the rank and file had of our agricultural program."[23]

In 1941 South Dakota state Democratic Party officials resumed these criticisms. They complained that the state's FSA leadership were all Republicans. "The entire set-up will continue to be a headache for the Roosevelt Administration until some of the switch-hitters are ousted," wrote E. H. Bremer. Former Democratic governor Thomas Berry protested the appointment of Ervin Trosin, a Republican, to an administrative position in the FSA: "It seems queer that in nearly all cases of this type, a Republican of the strongest background gets the job." Trosin was "a registered Republican, who had loudly and repeatedly denounced the New Deal and everything Democratic."[24] Despite ideological conflicts with the New Deal, FSA administrative and supervisory jobs may have been the only work around for these Republicans. This meant underpaid FSA staffers at the state and county level were often at odds with the policies of Democrats running the national rural rehabilitation agency. Therefore, the rural rehabilitation doctrine became garbled as it filtered down from liberals at the national level, to "conservative progressives" such as a Cal Ward at the state level, and finally to conservative Republicans at the local level.

The goals of the Resettlement Administration and FSA also changed with the severity of the drought and economic depression tormenting the country. Between 1935 and 1938 many in the USDA believed that even an economic recovery would not erase high rural unemployment. Therefore, the goal of the rural rehabilitation program was to engage as many people as possible. In addition, the FSA aspired to keep commercial agricultural production down while helping their clients furnish as much of their own subsistence as possible. However, this ran counter to both reason and agricultural trends in the plains. The idea of maximum employment with minimum production was contradictory. With good reason many plains farmers believed that the best chance for long-term success came from greater productivity while cutting operational costs through mechanization. A program to occupy as many farmers as possible and to produce a limited amount of marketable commodities challenged the expansionary entrepreneurial spirit that animated rural America.

Furthermore, the perceived goals of rural rehabilitation varied considerably. One social work investigator described rural rehabilitation as a far-reaching farm program. Rural rehabilitation, he wrote, worked as "a ladder which starts with very small steps of rehabilitation among people who are at the very bottom and which reaches up into the field of the transformation of tenant farmers into farm owners."[25] The Kansas Farmers Union agreed, claiming that the rural rehabilitation program was to "take care of that class of farmers who are unable to obtain financial support from private sources or existing farm agencies of the federal government."[26]

Others in the state were less hospitable. The *Topeka Capital* explained how regional director Cal Ward first had to deal with the emergency relief needs of plains farmers. "What to do with them after this emergency is over is a problem that will have to be met later." Ward noted that county committees approved loans based on the farmers' "experience, character, ability to conduct farming operations and also on the basis of need. . . .

These latter will have to be taken care of some way." The *Capital* reacted scornfully: "[I]n other words, just plain old relief, if there is no other way out. Ward cannot say it; but that is the fact just the same."[27]

Just what was rural rehabilitation? Nebraska's state RA director, L. A. White, interpreted rural rehabilitation as restoring farmers to their previous status by raising the farmers' standard of living up to local production and marketing conditions. Rather than relief, White considered rural rehabilitation "a business-like plan through which farmers are given a sound chance to work out their own problems." Tangible goals included working off debts, curbing losses, and getting the farmer off relief.[28]

In reality the rural rehabilitation approach, as practiced in the plains, attempted to move beyond relief. Specifically, rural rehabilitation was a program for loans and grants combined with technical assistance to reform farming and domestic practices. At the center of this was the farm and home management plan. Under this plan, according to the FSA, the farm family gained three essentials: self-sufficiency in food and livestock feed, a diverse farm enterprise for the commercial market, and farming practices that contribute to soil fertility. After formulating a plan, farmers received a loan payable in five years at 5 percent interest. With this financial and advisory assistance, the FSA planned to help the farm family "lay this foundation for a come-back."[29] However, most evidence suggests that farm families applied to the rural rehabilitation office for its easy credit, not for the supervisors' guidance nor the chance to convert their farm operation. Still, L. H. White described rural rehabilitation as more than just a loan. White said, "It is a step-by-step attack on the forces against which each farmer must fight," he announced. "The whole life of the farmer is brought into the plan and the work of each member of his family plays its part in the drama of restored independence: the risk of the venture is minimized and the security strengthened."[30]

The rural rehabilitation program's bureaucracy was set in place by 1935. Its first task was to save rural families on the brink of disaster. In the face of the punishing drought and heat of the 1930s, the goals of "independence" and "security" took a back seat to the immediate problem of sustaining a huge class of the rural poor. Between 1935 and 1938 the rural rehabilitation program dedicated itself to helping plains farm families simply survive. It had little choice. The problems of drought during these years forced the Resettlement Administration and the Farm Security Administration to concentrate on the emergency relief needs of the region. Distributing emergency feed and seed loans and grants was the priority. So any attempts to transform the farming operations of the rural lower-income groups waited behind providing subsistence for these families. Without feed and seed loans, these farm families had no income from their fields or livestock in the coming year, and they clearly needed help.

In North Dakota during the mid-1930s the RA and FSA operated primarily as rural relief agencies out of necessity. In early 1936 state director Howard Wood listed the types of families needing assistance: "stranded farm families" attempting to make a living on too few acres in semiarid parts of the state, those trying to cultivate land best suited for grazing, and foundering families requiring supervised credit. Furthermore, farm laborers and newly married couples needed aid and many "stranded industrial workers" comprised of unemployed railroad workers and coal miners living in rural areas desperately needed help.[31]

These groups made up a huge portion of rural North Dakota. Therefore, the rural rehabilitation program's clientele included those not normally considered indigent—and even those not normally considered true farmers. Wood's assistant director, Leonard Orvedal, recalled the enormous caseload of the early rural rehabilitation program in North Dakota. He cited one county with 1,800 farm families that received 1,814 grants. According to Orvedal, when a Washington staffer learned of this, "that threw [his] swivel chair into a tizzy so he came out here to investigate this." Orvedal explained the situation to the staffer by describing families living on the edge of town: "Whether they're part of the city or part of the country we don't know. Maybe they've got ten acres and a cow, but our policy is that we would rather feed ten families too many in any given county than fail to take care of a family that really needed it." Orvedal continued, "Welfare has got its hands full now taking care of the unemployed and the poor in the cities." He concluded, "Anyone who had a semblance of living off the soil we took 'em as a matter of principle and took responsibility for 'em."[32]

Under these conditions, the Resettlement Administration operated as a catchall rural welfare office. During the agency's year-and-a-half existence staffers worked frantically in North Dakota and the rest of the plains region not to transform farms but simply to keep them operating. Severe drought conditions in 1935 and 1936 placed the agency's work in the national spotlight. Farmers sent frantic letters, telegrams, and representatives to Washington DC and the Roosevelt administration pleading for aid. Federal relief officials opened their limited coffers to dispense between $14 to $17 a month to farmers in loans or grants to tide them over. The *Topeka Capital*, usually critical of the New Deal, celebrated in November 1935 that such aid "will offer a new lease on life to the farm families now facing the end of their row."[33] However, rather than a lease, this small stipend was only a temporary stay from indigence.

The Local Rural Rehabilitation Machinery

The administrative setup of the local rural rehabilitation program greatly shaped its performance. The farm supervisor, usually a male, acted as the head of the program in each county. It was his responsibility to collect and investigate applications for rural rehabilitation loans and grants and make debt and land tenure terms amenable to his clients. His other administrative duties included collecting loan payments and preventing client loan defaults and foreclosures. The supervisor also spent a great deal of time visiting clients to help plan farm operations and develop a leadership role among them. Finally, the FSA county offices promoted cooperative associations such as grain elevators or threshing operations, medical cooperatives, and neighborhood action groups.

Each farm supervisor was supposed to work with a county rural rehabilitation committee of three farm men or women. This committee counseled the supervisor on the suitability of loan and grant applicants, encouraged farm and home planning among clients, and fostered client involvement in community activities and cooperatives. According to the FSA, the purpose of the rural rehabilitation loan was to assist farm families on "economic size units which will provide a . . . satisfactory level of living and cash income sufficient to pay annual farm and home operating expenses, repay capital obligations, and allow the family to participate, within their capabilities, in the normal social, educational and economic activities of the community."[34]

The FSA guidelines dictated that the farm supervisor and the county committee approve loan or grant applicants and supervise their progress along very particular guidelines. For approval, clients had to have suitable land, secure tenure of this land, and, if they had debts, liabilities that could be scaled back. Finally, the client family had to cooperate with the rural re-habilitation program. This meant a commitment to producing home-grown fruits, vegetables, poultry, meat, and other livestock products for subsistence. In addition, clients were to supply their own pasture, forage, and grain for their livestock. They were also to generate enough cash in-come to repay old debts and handle cash farm and home expenses. Staffers expected clients to keep the farm and home account books up to date and repay FSA loans as stipulated by the FSA's terms.[35]

The FSA farm supervisor kept a vigilant eye over his clients' personal lives and the farms themselves. Couples, for example, might have to undergo a medical exam and investigations about their home life and their children's education. Following the applicant's approval for a loan or grant, the super-visor visited the farm family to monitor their progress. After surveying farm and farm home conditions and examining their budget, he filled out a farm management plan with the farm family. If the family had kept rec-ords, not always the case, the supervisor inspected the current operation, including their farm management practices, crop and livestock production and income, expenses, assets, and liabilities. Using this information, he made income projections for the coming year, suggested changes in the farming operation, and recommended further loans and grants when nec-essary. FSA guidelines constricted supervisors when offering credit to clients. Loans were only for rent payments, foundation and subsistence herds of livestock and draft animals, and land-clearing operations. Clients could also take out loans to buy farm machinery, household equipment, and fertilizer, and to construct small buildings and fences. Only when the client was threatened with seizure of essential property or foreclosure was FSA credit used to repay nonmortgage and unsecured debts.[36]

While the farm supervisor worked to buttress and improve operations for the field and farmyard, the home supervisor, invariably a woman, en-couraged better living conditions and greater self-sufficiency within the farm home. The home supervisor checked minute housekeeping details. She encouraged farm women to improve their standard of living through careful budgeting, self-sufficiency, bartering, and supplemental income. Home supervisors also advocated better living and sanitation standards by keeping the home in repair. Of these goals, the home supervisor stressed self-sufficiency most of all. Client families were expected to supply as much of their own clothing, soap, and even furniture as possible. Particularly important was self-sufficiency in food. To this end, the Holy Grail of the home supervisor was the pressure cooker. Supervisors encouraged clients to purchase a cooker with their FSA loans to can their own pork, vegetables, and fruit to cut expenses.[37]

Helen Buhler Ossman recalled her experience as a home supervisor between 1940 and 1943. Like many other home supervisors, she had a degree in home economics. Fresh out of the University of Kansas, Ossman became a home supervisor in Anderson County, Kansas. She and the county's farm supervisor visited client families twice annually. On an average day the pair visited two families in the morning and one in the afternoon, and wrote reports the remainder of the day. Ossman spent her time showing farmwives how to keep books, sew, use a pressure cooker, and grow vegetables. Together, the farm and home supervisors had a caseload of two hundred farm families.[38]

Some contemporary observers complained that the Resettlement Administration and Farm Security Administration treated their clients differently than private creditors would have. They charged that the rural rehabilitation program coddled clients, acted in an authoritarian manner, or held them back. This accusation is unfair and inaccurate. Rural rehabilitation clients were, for the most part, small-scale, lower-tier tenant farmers with sparse collateral. Most commercial banks, landowners, insurance companies, and government credit agencies would have closed their doors to them. Even if commercial credit institutions had lent these borderline farmers money, they would have closely dictated their operations. The primary difference between banks and the rural rehabilitation program was that the RA and FSA demanded that clients keep detailed accounts and that they diversify into livestock farms and semisubsistent homes. As noted, the goal of the rural rehabilitation program was not advancement or keeping up with the rising scale of agriculture. Rather, it was simply to allow the farm family to pay their work and living costs, reduce their debts, and maintain a viable standard of living until prosperity returned.

Rural Rehabilitation: A Fool's Errand?

By 1939 the Farm Security Administration, which succeeded the Resettlement Administration, finally focused its resources on rehabilitating its clients, rather than primarily providing them with relief. That year, for the first time, rural rehabilitation loans totaled more than FSA grants in the

Great Plains. The FSA now consciously steered itself away from the needy lower-income, small farmers to focus on borderline farmers who had a chance of rebuilding solvent farms. From this point on, the New Deal agency mostly lent to the more "substantial" farmers, those on the border-line between poverty and security. The result was that in the wetter, eastern plains farmers needed to own or rent at least 100 to 160 acres of land to qualify for a rural rehabilitation loan. Farmers in the semiarid plains had to cultivate at least 320 acres for FSA assistance. Other prerequisites for prospective FSA borrowers were sufficient machinery, livestock, additional farm labor, and a farm background. A huge group consisting of foreclosed farm families, tenants, retirees, unemployed workers, and young couples with few means had no chance for credit through the FSA.

For example, Clyde Abbott, a farm worker from Elkhart, Kansas, wrote Congressman Clifford Hope in 1941 that he had neither the farm machin-ery nor the credit to start his own operation. During summers, Abbott la-bored for "successful farmers." For the past three winters he worked for the WPA. Abbott complained that the FSA turned down his application for a loan, ostensibly because he lacked a longer-term tenant lease. The farm worker protested that it was impossible to acquire more stable rented land "due to the fact that the county is over run with suit case farmers or city farmers who merely scratch the ground to raise weeds and draw allotment checks." Representative Hope replied that the FSA required its clients to have longer-term leases, since the agency expected them to take several years to repay their loans. Hope closed sympathetically, "I realize what a man is up against out in our country in getting a farm these days."[39]

For lower-income ambitious farmers such as Clyde Abbott, the FSA's guidelines and Congressman Hope's empathy offered little solace. Unfor-tunately, the FSA possessed inadequate funds, was short on staff, and con-fronted a comparatively overpopulated countryside. Therefore, the agency wisely focused on helping the more promising group of farm families rather than trying to save the entire plains countryside. Faced with the agency's limitations, it made little sense to build or rebuild the smaller, unstable, and woefully undercapitalized farms which had little chance of surviving, much less keeping pace with the increasing scale of agriculture in the Great Plains.

Although the FSA's policy was logical, it quite possibly left 130,000 plains farm families, or nearly 600,000 people, dependent upon local re-

lief or the WPA. The loan program was an inadequate solution since it failed
to address the serious long-term problems that troubled the rural plains.
There still existed a great need for some kind of relief for lower-income
families in the American and plains countryside on the eve of World War
II. Pockets of rural poverty, consisting of hard-core low-income farm families, spotted the plains even as the rest of the nation gingerly entered an
economic recovery before Pearl Harbor.

The borderline farmers' progress under the rural rehabilitation program
presented a mixed record. An ideal approach to reforming plains farms
would resolve the primary problems of the borderline farmer: problem
tenancy, inadequate income, high debts, relatively small-scale farming,
economic and social instability, overdependence on relief, and the lack of
affordable credit. The program also would have to help borderline farmers
shift their farming operations to suit the market, environmental, and agricultural transformations of the interwar years.

There is no comprehensive portrait of the plains rural rehabilitation
client over the decade of the program. Therefore, the historian is left with
the task of piecing together short-term regional surveys to describe borderline farmers who sought government assistance. The most comprehensive
figures come from a 1938 FSA progress report for the plains which covered
eighteen hundred rural rehabilitation client farms in nearly every county
of the four-state area. The data indicates the place of these client families in
rural plains society. Eighty-six percent of the surveyed clientele were
renters, making rural rehabilitation a tenants' program. The average
client's farm had 317 acres. One-third of the land was in pasture, another
third used for cash crops, one-quarter used for feed crops for livestock,
with the remainder used for "other purposes."[40] This data fits precisely the
figures for plains tenant farmers during the 1930s. In 1935 plains tenant
farms averaged 313 acres, and a report on tenant farmers in Box Butte
County, Nebraska, portrayed renters with a similar disposition of land in
crops and pasture.[41] The surveyed clients' annual income was just at the
poverty line; even then they required a rural rehabilitation loan to raise
their income to this level. Certainly many clients' income slipped below
the $1,000 a year needed for a sufficient standard of living for rural Amer-
ica during the 1930s.

To place clients on their feet rural rehabilitation staffers made farm subsistence a major tenet of the plan. This meant avoiding tractor farming.

Plains farm supervisors informed their clients that draft animals, rather than machine power, were best suited for their operations. FSA supervisors made loans for clients to purchase tractors and other power machinery only when they could not cooperatively use a tractor with other farmers. In addition, farm supervisors steered clients away from the volatile grain commodities market and toward cattle raising. Clients could grow cash crops only when they first had sufficient livestock feed on the farm to contribute to a "well-balanced program."[42]

The FSA in the plains was exacting about what constituted this "well-balanced program." The agency promoted "the advisability of changing from cash crops to a livestock program including a variety of cash crops, permanent grass and feed crops." The idea was to make the client's farm more stable in the face of low rainfall and feeble market prices for cash crops.[43] This marked a retreat from the commercial wheat, corn, and hog-centered farming that dominated the region.

The agency offered an alternative based on subsistence crop farming augmented by cash from the sale of livestock and livestock products. Across the plains FSA supervisors insisted that clients raise drought-resistant feed crops such as sorghum as part of their rehabilitation. They also maintained that farmers who built up "subsistence herds" of livestock could live off homegrown meat, eggs, and dairy products and spend less on groceries. The FSA survey approvingly reported that between 1936 and 1937 farm clients substantially enlarged their stock of workhorses, dairy cattle, bulls, and poultry.

Clients found that to receive their loans they had to forsake grain commodity markets. In 1939 half the money lent in the region went to purchase livestock, usually for breeding and draft work.[44] Those clients who steered away from small-scale livestock growing quickly learned that their only source of credit disappeared. For example, in early 1940 one FSA client with a 160-acre farm complained to Congressman Clifford Hope that he had been denied further credit. Regional FSA director Cal Ward explained to Hope that the McCracken, Kansas, farmer had changed his operation when he began renting a 536-acre farm. FSA staffers rejected the loan because, in their opinion, the farmer lacked the resources to run such a large property. Furthermore, the operation was "highly speculative" given that the farmer planted two hundred acres of wheat for his income. Ward explained that the FSA made loans to establish diversified farms using work stock to cut operating costs and "insure a stable income." Congressman Hope suggested the farmer apply for a loan at a commercial bank.[45] It was a futile piece of advice, since most FSA clients were unable to secure credit from banks in the first place.

The rural rehabilitation program wasn't asking its clients to do anything new or out of the ordinary. Most of them had livestock, practiced some self-sufficiency, and used horses rather than tractor power. What was new for clients who liked farming their own way was being told how to work and live their lives. Now tenant farmers had another boss in addition to their landlord and creditors—their rural rehabilitation supervisor.

Clients found other drawbacks to the rural rehabilitation plan. The FSA's strategy of encouraging semisubsistent livestock growing probably curbed their short-term income. Since clients partially withdrew from commodity markets, they missed the opportunity to profit from them as other farmers did. In South Dakota, for example, farm supervisors insisted that clients secure three-quarters of their income from livestock and livestock products. The FSA attributed a drop in clients' income there to the staffers' pressure on them to keep a foundation herd rather than sell it.[46]

FSA clients were also unable to increase the scale of their operations. Since they were committed to stability at the expense of productivity, clients were fighting the trends of the time. Progressive farmers cut labor expenses through mechanization and cultivating more acres efficiently and productively. If the commodity markets remained flat, the growing farm entrepreneurs fell back on federal commodity payments. If more robust markets returned, as they did during World War II, these industrious farmers could reap the benefits of higher agricultural prices. The rural rehabilitation program, in contrast, committed farmers to small-scale, labor-intensive production.

Finally, plains FSA clients couldn't survive without direct relief in the form of FSA grants, and they became dependent on these grants once they began the program. Between 1934 and 1942 plains clients received $59 million in grants.[47] In Nebraska the FSA attributed this reliance on grants to higher home expenses. Years of drought took their toll as families had to purchase their food rather than grow it. Also, after years of doing without, many families needed grants to replenish their worn-out clothing.[48] Many New Dealers feared clients would become addicted to federal aid. FSA director Will Alexander countered, however, that distributing grants along with loans to new clients was more realistic early in the rehabilitation process. The alternative was saddling the clients with large debts from the beginning.[49] Yet the FSA's goal in distributing direct relief to its clients left its leadership uneasy. In 1938 the Kansas FSA director presented a portrait of clients' crop diversification, increased subsistence production and living standards, and the adjustment of debts. However, he admitted a "dark spot" in the continual demand for grants from clients.[50]

Rural Rehabilitation Clients: Businessmen or Beggars?

By contemporary standards, the rural rehabilitation program was a largesse for the Great Plains. From its beginning in 1935 until the end of World War II, plains farmers received $106 million in rural rehabilitation loans.[51] Yet two questions dogged the rural rehabilitation program at the national and regional levels. The first was whether farm supervisors should follow commercial banking principles and pressure clients to repay their loans. The second question concerned whether funds should be devoted to a broad-based clientele (as did the rural rehabilitation program) or to limited programs such as the tenant purchase program discussed below. The Resettlement Administration and the FSA responded ambivalently to the first question. Rural rehabilitation loans were advanced under liberal conditions at moderate interest rates but were still closely tracked for repayment. In response to the second question, the FSA focused on both a mass group of rural rehabilitation borrowers and a select group of tenants who qualified for loans to purchase farms.

For public consumption, proponents of the rural rehabilitation loans painted them as prudent investments in borderline farmers who had the integrity and reputation, but not the security, to guarantee a commercial farm loan. Cal Ward presented the loans as complementing those of regular banks. He praised the program for adjusting clients' debts so they could be repaid and for allowing worthy farmers access to credit. The FSA, Ward said, also reformed farm practices to put the operation on a "paying basis" and rehabilitated the client for the good of the commercial life of a community. Ward explained that rehabilitation loans filled a credit void for borderline farmers. He pointed out during the 1920s a farmer went to a bank and received a "character loan" on the basis of the farmer's reputation rather than collateral. Since such loans were no longer feasible, rehabilitation clients turned to the federal government for credit.[52]

Laird Dean, president of the Kansas Bankers Association, reiterated this at a conference of farm supervisors in Salina in 1939. Dean, described as a conservative Republican, defended the loans. He praised the mission of the Farm Security Administration, but not before freeing commercial bankers from any obligation to small farmers. "The banker's job is not that of rehabilitation. Our responsibility is to our depositors. We must safeguard and invest their money," he said. "Rehabilitation of hard-hit farmers is society's job and can be done only by the government." The *Salina Journal* applauded Dean's comments and commended the FSA which, in addition to loans, lent courage and boosted the morale of its clients. Such loans sought both to improve the farmer and reform the farmer "mentally as well as in a material way." The alternative was to abandon farmers "who, if left to drift, would become permanent public charges on city relief roles and prey for agitation."[53] Perhaps the newspaper's editor recalled attempts by the United Farmer's League, affiliated with the Communist Party, to recruit farmers. Or he feared the potential for tenant farmers to organize along the lines of urban labor unions.

Yet there were also critics of liberalized agricultural credit. They claimed easier credit was a bad risk and burdened farmers with debts they could not repay. For example, South Dakota's State Planning Board warned that loans extended without adequate collateral could create "an anti-collection psychology" among recipients. The board specifically cautioned against making credit available to incompetent farmers or to those on land unsuitable for their operations.[54] Others within the Department of Agriculture, such as the Farm Credit Administration (FCA), questioned the usefulness of the rehabilitation loans. Staffers at the FCA, which refinanced mortgages and gave operational loans to more substantial farmers, criticized the FSA for offering credit to undeserving farmers who would use the funds improperly.[55]

Agricultural economist John D. Black acknowledged the deficiencies of rural rehabilitation's lending policies. In a 1943 report, Black noted the changing nature under which the rehabilitation loans were made. During the mid-1930s the loans were meant, in Black's words, "to rebuild families that thus far have made a mess of their lives." With the approach of World War II and the return of rains and somewhat higher farm prices, however, the loans were meant to expand farm production and boost income of farm families whose lack of resources impaired their success. Black noted that while both goals were important they countered each other. Black recommended splitting the rural rehabilitation program into two programs: one for those farmers who could plausibly increase the scale of operations, or borderline farmers, and another program for the "truly underprivileged farm families."⁵⁶

The above statements came from self-serving sources. Men such as Cal Ward and the Kansas banker supported the rural rehabilitation loans as prudent investments for their own reasons. On the other hand, experts at the regional and national level doubted the usefulness of making loans a commercial bank would decline. Yet, as John Black would have noted, these arguments ignored the complex composition of the FSA's clientele and the plains countryside. Thousands of the region's farm families desperately needed RA and FSA grants and loans in drought-stricken areas. Thousands more made up a sector of hard-core rural poor. These were lower-income families with few resources, living in poorer farming regions, who had low living standards before the Great Depression. Then there were the borderline farmers who had the resources and the ability to expand their production if they had the resources, credit, and luck to stick out the bad years.

Naturally, critics debated just who the clients were and how the rural rehabilitation program should treat them. A dilemma for the Resettlement Administration and Farm Security Administration was to define their clientele. Were they indigent farmers on relief or were they up-and-coming entrepreneurs? One approach treated the clientele as victims who had to be aided and regulated closely. As such, the FSA treated the farmer requesting a loan the same way the local relief board treated an applicant. Prospective borrowers filled out lengthy forms and submitted to probing interviews to determine their need and worthiness for the loan. Every effort was made to weed out the "undeserving" from the "deserving" poor. Supervisors subjected both rural rehabilitation and relief recipients to investigations to ensure they used government funds appropriately. The FSA took this method a step further by inserting their farm supervisors in the business. Clients could not make major changes in their operation without the supervisor's approval. They had to submit checks drawn on FSA funds to supervisors to be cosigned. This especially bothered farmers who treasured their freedom to run their operations as they pleased.

On the other hand, the official philosophy of the rural rehabilitation program was to trust and encourage clients with loans and advice. Supervisors then measured their progress by the business standards of the time. This meant FSA staff expected clients to increase their livestock production and income, repay their loans, turn a profit, then position themselves for further growth. This combination of social work and business values meant the FSA treated clients both as supplicants and as small, commercially oriented entrepreneurs. Because rural rehabilitation clients were judged as rural businessmen, an essential gauge of the rural rehabilitation loans was repaying loans on time, which they did. By 1946 plains clients had repaid 82 percent of the total loan amount. This compares favorably to the 78 percent repaid on the $1 billion in rural rehabilitation loans distributed throughout the nation.[57]

Harold Clingerman, who managed his bank's rented lands in central Nebraska, noted his problems with tenants who were FSA clients. The agency's county office had to approve clients' farm operations first, even down to the slaughter of a pig for home use. Although Clingerman's bank didn't discriminate against FSA clients as tenants, he found that clients had brought a "third partner" onto their farms, and that the FSA had first claim on loan repayments. Locals called clients "rehabbers . . . with just a tinge of condescension." Clingerman too was confused whether the FSA "was a bank with a loan portfolio of questionable security or a social agency dedicated to welfare aid for its clients, some days it seemed to be one, on other days the other."[58]

The Resettlement Administration and the FSA embraced one program that had a huge and positive impact on debt-ridden farmers in the plains. The Farm Debt Adjustment Program sponsored committees at the county level. These committees, made up of five local representatives, counseled farmers and their creditors to accommodate farmers' debt loads with their ability to make repayments. These committees had no legal authority, but they succeeded in forestalling foreclosures against insolvent farmers by banks, local businesses, and individuals through negotiations. Between 1935 and mid-1941, committees throughout the plains adjusted the debt load of nearly twenty-five thousand farmers, a third of whom were not FSA clients. In the average case, the committee cut the farmers' average debt of $3,440 by $1,150, or 33 percent.[59] In most circumstances this saved creditors from expensive foreclosure procedures with little in judgment payments.

Despite such financial and managerial assistance, however, rural rehabilitation clients offered a mixed short-term record for repaying their debts
from the FSA and other parties. Individual case histories from eastern Nebraska's FSA files illustrate clients' varying performances in 1940. James
Kluge of Cedar County was paying off his loans on schedule while increasing his net worth fivefold between 1938 and early 1940. Burt Wasser of
Madison, Nebraska, remitted his FSA loans within two years. In addition,
he was keeping up with the payments on his father's mortgaged farm. But
FSA borrowers Leo and Paulette Knapp were less successful. The Knapp
farm, like others in Thayer County, had been hit by drought since 1937,
and the Knapps covered only the interest on their loans. They required several supplemental FSA loans to meet feed and seed expenses. Thayer
County's farm and home supervisors reported that the Knapps' operation
needed improvements and that the family exhibited too much autonomy.
The supervisors closed their narrative: "The family takes the suggestions
of the farm and home supervisors whenever they feel that those suggestions will improve their situation." Still, the FSA office recommend-ed
another five years of rehabilitation for the Knapps.[60]

A five-year period is a long time to expect clients to remain under the
close supervision of a government agent. However, it is not a long time to
plan the future of a farm. This reveals the schizophrenic nature of the
rural rehabilitation program. The plains FSA was attempting to fuse principles of social work and agricultural extension with entrepreneurial expectations to transform their businesses. Under such a divergent program,
borrowers were succeeding as clients. However, their steady appetite for
supplemental loans and grants and technical support suggests they were
failing as businessmen and women.

Rural Rehabilitation and Farm Tenancy

The Resettlement Administration and the Farm Security Administration
both attempted to help a broad-based group of farmers make incremental
improvements in their farm operation and standard of living through the
rural rehabilitation program. In addition, the FSA sought to convert a select
group of tenants into farm owners. As noted, nearly all the rural rehabilitation clients in Kansas, Nebraska, and the Dakotas rented their land. Observers correctly associated problem farm tenancy in the plains with diminished income, poor living conditions, and social instability. Therefore,

the attempt to correct the problems of farm tenancy among plains border-line farmers deserves close attention. Correcting the long-term causes of problem farm tenancy was beyond the scope of rural rehabilitation. Instead, the New Deal program chose to address the results of these problems. These difficulties included short-term leases, a dependency on cash rents and single cash-crop agriculture, and the failure to reward improvements made by tenants on rented land.[61]

While farm planners recognized the long-term trends at work in tenancy, they focused on improving the tenants' leasing terms as the most plausible reform. The Resettlement Administration studied dozens of written farm leases from around the country and found that short-term leases with little encouragement for soil conservation were the norm. The agency observed that major investors in farmland such as insurance companies demanded one-year leases in order to regain immediate possession in case they found a buyer for the land. In addition, tenants were rarely compensated for improving soil fertility through crop rotation, planting soil-building crops, and using fertilizer.[62] To rectify these problems, the FSA attempted to improve lease terms for its tenant clients. To this end, the agency introduced a five-year "flexible farm lease" to match the duration of the rural rehabilitation program for individual clients. This lease gave the landlord the right to terminate the agreement at the end of the year with written notice and provided for compensation for tenant improvements and arbitration for disagreements between the landlord and tenant.[63] Though these terms were laudable, they were difficult to implement.

To fully understand the relationship between tenancy and rural rehabilitation, the USDA's Bureau of Agricultural Economics conducted interviews with FSA county supervisors in the Great Plains in fall 1940. In particular, agricultural economists Elco Greenshields and J. M. Stensaas focused on the impact of the rural rehabilitation program on their clients' tenant conditions in the region. Greenshields and Stensaas found the leading issues were improving rental terms and farm operations.[64] However, farm supervisors throughout the plains related the difficulty of setting up more favorable lease terms for FSA clients. Supervisors attempting to implement five-year leases found tenants had little leverage with which to compel landlords to improve their leasing arrangements.

Many times tenant clients lost their lease because of local conditions. A North Dakota farm supervisor pointed out that if the tenant disagreed with the terms of the lease, the landlord could tell the tenant, "Take it if you want it . . . or someone else will." Most renters had to accept a one-year lease. A South Dakota farm supervisor noted that some landlords reserved the right to cancel the agreement if the rural rehabilitation client failed to secure an FSA loan for that year. Many times clients had signed an initial one-year lease and simply continued the terms into the following years unless the landlord ended the agreement. A Kansas farm supervisor remarked that landowners rejected the FSA's long-term flexible farm lease. Even though the FSA leases had an annual termination clause, landlords disliked being tied to such an agreement. In Nebraska, tenants and landlords had a free hand in determining term length, where the agreements were only for one year, beginning on March 1. Under these circumstances, the FSA supervisor demanded that the short-term lease for clients be automatically renewable.[65]

Clients had to contend with more than the length of tenure on rented land. Shortages of rentable land throughout the plains gave landlords the advantage in setting rents. As noted, tenants often paid their landlords through a combined share of the crop and cash rent. While rental shares, established by custom, were fairly stable, tenants usually had to bid for the cash rent. This drove the price as high as $400 for 160 acres of farmland. Although one supervisor attempted to cut the cash rent for farms, he failed. In the face of higher cash rents, the supervisor had to allow clients to sign expensive renewal leases, since the alternative was for the client to have to move again, an expensive proposition. Similar situations existed in North and South Dakota. With an excess of tenants, landowners were free to set their cash rents.[66] Clearly, the plains had too many tenants trying to farm.

Landlords saw little need to invest in the rented farm. To counteract this, the FSA agent occasionally acted as a mediator to encourage the owner to furnish the materials and labor to renters for refurbishing the farm. One Kansas farm supervisor considered the improvements on rented clients' farms as "relatively poor," and the chief disagreement between tenants and landlords arose over reimbursement for improvements. Because of this, FSA clients received loans for only small improvements on rented farms, such as building henhouses or garden fencing.[67]

Since many farm observers considered farm tenancy as the enemy of soil conservation, the FSA encouraged clients to raise soil-building feed crops such as alfalfa, and to pay for it through share rental rather than cash rental.

However, it was difficult to convince landlords to share rent on such feed crops.[68] Landlords, particularly retired ones, depended on cash rents and the sale of their share of grains, such as wheat, to support themselves. In a depressed economy, the landlords' needs were immediate and dire. Therefore, it was difficult to convince them to commit to a soil-building approach, even when such a program enhanced the land's long-term productivity.

Finally, farm supervisors recognized that clients still using workhorses were at a disadvantage in the mechanizing plains countryside. More substantial farmers used tractors to cultivate larger parcels of land that were once part of rented farms. In South Dakota, insurance companies maintained the fencing and buildings on one property while tearing down those on another. They did this, remarked one farm supervisor, "because they can get just as much rent from two farms with one set of improvements as they can with both of the farms improved." He noted that "whenever a set of buildings is destroyed there is little likelihood that they will be replaced by another." In addition, insurance companies refused to rent land to tenants without tractors. Some landowners even declined to lease their property to farmers with mortgaged machinery or to FSA clients.[69]

Problem farm tenancy reflected the troubles of the rural plains in the 1930s. Many tenant farmers were undercapitalized, underproductive, and relatively inefficient. As renters, FSA clients were in a tenuous position. Not only did they make insufficient income, they also had little geographic stability and diminished opportunity to renovate their farm and practice soil conservation. Finally, since many of these tenants lacked tractor power, they couldn't cultivate the large acreages they needed in this age of large-scale agriculture. Markets set the conditions of plains farm tenancy before World War II, not the FSA nor the borderline farmer. Landowners, often short of cash, wanted as much from their land as possible. Therefore, they invested as little in their property as possible, encouraged cash grain farming and the use of tractors to increase their share, and refused longer-term leases in case a more profitable tenant or a buyer came along. Because of the glut of potential tenants, the landlords were free to choose among tenants and to set the leasing terms.

The Tenant Purchase Program in the Plains

One possible solution to the shortage of land and the unfavorable leasing terms for rural rehabilitation clients was to finance a farm through the

FSA's tenant purchase program. The drawbacks of problem farm tenancy, such as low income, soil erosion, social instability, and poor living conditions, became national issues in the mid-1930s. In response, in November 1936 President Roosevelt created the Special Committee on Farm Tenancy, chaired by Secretary of Agriculture Wallace. The resulting committee report on farm tenancy in America led to the Bankhead-Jones Act, intended to turn land renters into owners. A review of the controversy surrounding this act reveals the reasoning behind rural rehabilitation as well as the early support and opposition for the program. The debate in Congress over a national tenant purchase program also marked the only time a rural rehabilitation–affiliated program passed through formal legislative approval.

Up to 1937 the repercussions of the New Deal on farm tenants were mixed. On one hand, many renters collected relief and other aid from new government programs. Tenants such as Ray and Gladys Gist of South Dakota benefited from several government programs issuing veteran's payments, feed and seed loans, soil conservation payments, relief work, and from wheat and hog reduction programs.[70] On the other hand, larger landowners probably benefited more from the AAA, which paid farmers to take land out of production. As C. F. Parsons of Sterling, Kansas, complained, "if a man owns a few thousand acres the government will give him a few thousand dollars but if you haven't got the land they will give you a job on the W.P.A. if you are lucky."[71]

To make the Bankhead-Jones Farm Tenancy Act of 1937 more palatable to local conditions, Congress designed the tenant purchase program to fit the values of rural neighborhoods. Local committees bought farmland with government funds, then sold it to tenants with financing at low interest rates. The local committee nominated eligible renters, giving preference to established tenants with livestock and savings for a down payment. The bill drew supporters and opponents from around the plains, yet it had broad-based support. The Great Plains was politically and fiscally conservative; however, the region's government officials and the press favored a prolonged government obligation to convert renters into owners. N. E. Hansen, head of the horticultural department at South Dakota State College, advocated a surprising plan for a federal takeover of much of the countryside through "State Capitalism." Hansen stated simply that "federal ownership of land is the Master Key to the Unemployment, Farm, and

Labor problems."[72] Congressman Charles Binderup, a Democrat from Nebraska, echoed these sentiments. He called for a twenty-year government-held mortgage on tenant purchase properties with no payments during the first five years. Such a proposal, he declared, would save the tenants on taxes and fend off speculators interested in buying the land.[73]

The *Omaha World-Herald* endorsed Franklin Roosevelt in 1932 but deserted him by his 1936 reelection bid. Still, it backed a tenant purchase program in spring 1938. The newspaper feared a rural America "possessed by a land holding class living off the labor of farmers receiving little more than subsistence," as in the South. Instead of a new government agency handling tenant purchases, the newspaper supported loans through the federal land bank system. Still, the newspaper admonished, loans "can't do anything for the man with no other capital than his investment in live stock and farm machinery."[74]

The total cost of acquiring the land for American tenants varied enormously according to the source. A spokesman for the Farm Holiday Association from South Dakota suggested it would require $500 million. N. E. Hansen speculated that $10 billion would be needed for government purchases of the land for sale to tenants.[75] Congressman Clifford Hope, a Republican from Kansas, estimated it would cost $14 billion, at $5,000 a farm, to make the farm renter an owner—politically an impossible amount. However, Hope was convinced initially that a scaled-back tenant purchase program could serve as a demonstration project.[76]

Responding to the tenant purchase bill, farmers and nonfarmers throughout the plains flooded Washington with earnest requests for land and money. Earl Singleton of Hutchinson, Kansas, asked for aid to buy some land adjoining his father's property in Reno County, Kansas. A dentist in Columbus, Nebraska, like other small businessmen in plains towns, had lost money investing in farmland. "I want to get out of the farm game," he explained. He had a tenant, "a very fine fellow [who] wants to buy it providing he could get in on this new finance plan."[77]

There were those, however, who opposed a federal plan to convert the tenant into an owner. Anton Odvarka of Nebraska complained to President Roosevelt that tenants already had the better part of the lease. Landlords, not renters, Odvarka stated, had to pay the insurance, taxes, repairs, and so forth, even when there was a crop failure and no farm income. Odvarka claimed that farmers who inherited land before World War I often squandered it and now were the ones in debt and on relief. The Nebraskan also asked valid questions about the place of these new owners in the plains countryside. "You propose to establish tenants on their own farms," Od-varka wrote. "How long do you suppose they will hold them? Or will they ever pay taxes, interest and other expenses falling to owners?" He recommended thrift and hard work as the solution.[78]

The most powerful farm group in the country, the American Farm Bureau Federation, was skeptical of a tenant purchase program. Even though national Farm Bureau President Ed O'Neal of Alabama was on the Farm Tenancy Committee, he declined to support the tenant purchase plan. Will Alexander claimed that O'Neal was indifferent to the problem of farm tenancy and doubted (for unspecified reasons) whether renters could ever become true farmers.[79] The Farm Bureau cautioned Congress that it opposed "sour loans" for marginal land or creating a new agency or government-sponsored cooperatives to handle the loans. A year later, commenting on the Bankhead-Jones bill, the Farm Bureau stated that higher farm income through healthy commodities markets was the true solution to farm tenancy in America. It also reiterated that the tenant purchase program be confined "to worthy young men . . . of demonstrated ability, farm background, and moral worth." As a safeguard, the Farm Bureau recommended a probationary period for the loans. Furthermore, after paying one-quarter of the mortgage, the tenant purchaser should have the title to the farm.[80] The Farm Bureau gave its lukewarm support partly to bolster its reputation. The Bureau represented the interests of larger commercial farmers raising a select number of commodities and had close ties to New Deal crop control and mortgage relief agencies. Therefore, for the farm organization to publicly oppose this popular initiative on behalf of smaller farmers would have been politically unwise.

The Farmers Union sent out mixed messages on federal tenant purchases. Most of the farm group leaders at the national and plains levels seemed to support the plan. On the other hand, during World War II the Nebraska branch of the organization lambasted the tenant purchase program. In 1943 the *Nebraska Union Farmer* ignored the predicaments attending farm tenancy and claimed that the issue was in truth a "land problem." Tenant purchase loans were not the real solution to the cycle of land booms, speculation, and "gathering of unearned increment" that troubled the plains. In fact, the state's Farmers Union head asserted that the tenant purchase program, if allowed to grow, would worsen the tenants' situation.[81]

Congressman Clifford Hope of Kansas was also incredulous about the plan, but for different reasons. A ranking Republican member of the House Agricultural Committee, Hope took an active role during House hearings on the Bankhead-Jones bill. Early in 1937 Secretary Wallace recognized that any funding for a tenant purchase program would help only a few farm families. Wallace explained, "It is far more significant to raise several hundred families a few degrees [than to completely] rehabilitate a small percentage."[82] Congressman Hope was troubled by this notion. Although he was sympathetic to the problems of tenants throughout the nation, he expressed his doubts whether the $50 million proposed for the program would make a dent in farm tenancy. "As far as its being an adequate solution to the problem is concerned, it will be like trying to fight a regiment of soldiers with a pop-gun." Even at a low purchase price of $4,000 a farm, Hope estimated that the original budget would purchase only one-third of the new tenant farms created *annually* in America.[83] Hope summed up his frustrations during the hearings. He was bothered by "the apparent futility of trying to attack this great problem in such a meager way. . . . What I mean is, should we not start on a basis of perhaps not trying to make landlords out of these tenants, but trying to make conditions better for them as tenants?" Then the Kansan concluded, "Wouldn't it be more helpful and wouldn't the money go much further if you spent it purely on rehabilitation of a larger group rather than in establishing a comparatively small number on their own farms?"[84]

During hearings on the bill, Congressman Hope questioned FSA head Will Alexander on this point, since the agency would be running both the rural rehabilitation and the tenant purchase programs. Alexander vacillated on picking favorites but admitted that if it were a choice between spending on a rural rehabilitation or a tenant purchase program, he would choose the former. However, he diplomatically stated his desire that both programs be funded.[85]

In July 1937 the Bankhead-Jones Farm Tenancy Act became law and Congress set aside funding for a tenant purchase program. The Farm Security Administration was established to administer the act and the rural rehabilitation program. A month later, Hope noted that little had been done to move the tenant purchase program forward. In the meantime, he had received many letters from constituents seeking a tenant loan.[86]

The resulting tenant purchase program had its strengths and weaknesses. On one hand, the FSA was able to lend nearly $20 million to plains FSA clients to purchase their own farms by the end of World War II. Tenant purchase clients in the region repaid on their loans fairly quickly. By 1943 they had paid off one-third of the principle on their loans, which averaged $8,333 per farmer. Within three years, plains clients remitted 45 percent of the principle on these loans. This compares favorably to the repayment rate of 38 percent for the United States as a whole.[87] The tenant purchase program eventually made sixteen hundred loans for farms averaging 361 acres in size in the Great Plains, and only 4 percent of these loans were delinquent.[88]

Tenant purchase clients made progress in more ways than their debt repayment. In 1940 Dakota newspapers praised the positive impact of the loans on former renters. In South Dakota, the *Sioux Falls Argus Leader* lauded a $12,000 tenant purchase loan for the Singer farm of Elk Point. With the credit, Andy Singer renovated a dilapidated farmyard and increased his farm yields on his 160-acre operation. Aided by the FSA, Singer was "launching a new battle on the old farm frontier."[89]

The *Bismarck Tribune* pointed out that tenant purchase farms were to be "family sized." That is, they were to be small enough for a family to supply its own labor. Yet it was to be large enough so that the family could make payments on the loan as well as pay taxes, maintenance, insurance, and provide for an adequate standard of living. One client, Eddie Schauer of Sterling, North Dakota, had a $4,843, forty-year mortgage with the FSA. The tenant purchase program existed for farmers like Schauer, according to the *Tribune*. The loan provided him "with the boost he needs to make him an owner independent and prosperous on his own land and an asset to his community and state."[90]

These loans were a boon to those fortunate enough to have them. However, the tenant purchase program was able to take on only a small fraction of the applicant farmers. While the *Bismarck Tribune* article cited above applauded the program, it noted that the committee approved only four of the ninety tenant purchase applications sent to the county office. These numbers reflected the regional record. Between 1938 and 1940 plains farm supervisors received 7,400 requests for tenant purchase loans. During this time, the FSA lent to only 550 farmers an average of $8,200 to buy a farm, which meant that only 7 percent of the applicants secured the credit and a farm.[91] Obviously, this discouraged the thousands of plains tenants desiring to buy their own property.

Andy Singer and Eddie Schauer were quite favored indeed to qualify for a tenant purchase loan to buy a farm of their own. The two were among the "elite" of FSA clients who received a large loan with perhaps long-term security. However, would they have the opportunity to increase the scale of their operations and keep their farms going? And what part did the RA and FSA play in such aspirations? The FSA's tenant purchase program was the agency's opportunity to transform a small proportion of plains farm renters into owners. While it achieved that goal quickly and decisively for a few clients, the task of making large loans to tens of thousands of American tenant farmers was too vast a project to support. The tenant purchase program's defects reflected the mixed feelings of the farm sector toward the true solution to rural America's problems. Despite the broad initial backing for a plan to assist problem renters, the tenant purchase program never went beyond a showcase project. At the national and regional level, representative voices expressed their doubts about the tenant purchase program. The Farm Bureau advanced a market-driven solution while Congressman Hope saw rural rehabilitation as the answer.

The real manager of the rural rehabilitation program was the hard environment of the 1930s. Throughout its inception, development, and operation, the plains rural rehabilitation program faced a conflict between emerging relief demands, short-term desires for profits, and long-term planning needs. At the same time, it was caught amid the clash between government and commercial solutions. The Great Plains was eager for federal assistance, and after a confused launching of the rural rehabilitation program, federal dollars flowed into the region. However, the region was selective in the kind of government assistance it welcomed. Rural rehabilitation loans to prop up local borderline farmers were justified as "business propositions." Tenant purchase loans, when economically feasible, were praised for putting renters on the path to farm ownership.

Rural rehabilitation clients faced real constraints on their farm operations when they accepted grants and loans. The rehabilitation goals for "balanced farming" put them at odds with the entrepreneurial ethos of increasing scale and maximizing income. But up until 1939, assistance was palatable to Americans when economic and drought emergencies mandated aid for farmers. With their relative shortage of capital, income, and land, many borderline farmers couldn't have enlarged their operations anyway. Their primary goal, as well as that of the FSA, was to simply keep their farms going and their families together. Clients such as Mrs. Hannon of Kansas saw the FSA as a safety net that checked her family's descent into insolvency. Yet when the farm economy began to recover in the early 1940s, the FSA's safety net seemed to some borderline farmers more like a trap keeping them in place. The entrepreneurial values during these prosperous times rebelled against the FSA's "balanced farming" tenets. The government program was geared toward a balanced income, rather than an increasing one, which conflicted with the commercial goals of many clients. This portrait of the plains rural rehabilitation program is necessarily limited. Its goals and operations at the grassroots level merit a closer look.

5

Farming in Place

The New Deal at the Grass Roots

The vast portrait of history consists of innumerable imprints from lesser individuals. Until now, we have followed farm insolvency in the huge Great Plains region through the early 1940s and focused on farm families in Kansas, Nebraska, North Dakota, and South Dakota, many of whom were tenants. The federal government made innovative yet sometimes flawed efforts to provide relief and to reform these farm families. The rural rehabilitation programs of the Resettlement Administration (RA) and the Farm Security Administration (FSA) arose in order to offer grants, loans, and advice to them. So far we have explored a countryside writ large to examine a large group of borderline farmers, those grossing between $500 and $1,000 annually during the 1930s. These were men and women who aspired to middle-income security. Unfortunately, the drought and economic depression of the decade drove them to the margins of insolvency. We have examined the reactions from the rural plains, the state house, the White House, and Capitol Hill to assisting such farmers. Now it is time to look at a miniature portrait of farm poverty and rehabilitation at the grass-roots level of two plains counties during the 1930s and early 1940s.

The era between the two world wars offers an exceptional opportunity to observe daily life and government activism writ small. President Roosevelt and his New Deal were popular precisely because they brought government action to the county level. President Hoover sought to spark a recovery from the upper economic and political echelons of America. The New Deal progressed beyond Hoover's framework of limited intervention into the economy. Correspondingly, the Resettlement Administration started to relieve the public's distress by establishing a presence at the lower levels of

government administration. County-level activities, such as the rural rehabilitation program, were the primary efforts against the Great Depression.
Furthermore, the county was the lowest administrative unit for most published USDA and census reports. Therefore, studying the county level al-
lows us to more closely examine and evaluate government efforts to
change conditions at the local level in the plains countryside. The goal of
this chapter is to assess the impact this farm program had on its clientele.
For this purpose, I selected two plains counties for a grassroots study of
the rural rehabilitation program.

When prospective clients came into the local rural rehabilitation office for
assistance, the farm supervisor invariably handed them application forms.
Farmers grumbled about the mounds of paperwork they had to fill out
(they still do) before receiving a grant or a loan. The clients' pain is the historian's gain, since many rural rehabilitation loan case files have been
saved for selected counties throughout the United States. These files con-
tain a wealth of information about clients' farming operations and their
economic and social characteristics, as well as their progress and attitudes
toward FSA goals. In this chapter I compare two counties' rehabilitation experiences to illustrate factors and problems in rehabilitating plains borderline farmers. Case studies should have more than one subject in order to
flesh out differences between the norm and the anomaly. For this purpose,
Barnes County, North Dakota, and Coffey County, Kansas, were chosen to
represent the varying rural rehabilitation experience in the Great Plains
(see map).

Barnes County, North Dakota, was representative of the plains experience during the late 1930s and early 1940s for several reasons. By plains
standards, Barnes County was average or close to average in farm size,
gross income, the value of the farm operation, mortgage debt, agricultural
productivity, and land values. In contrast, several attributes of Coffey
County, Kansas, suggest rural rehabilitation of its clients would have been
difficult. Overall, Coffey County farmers had comparatively lower income
and productivity, low property values, lower managerial ability, and soils
unsuitable for small-scale agriculture. These factors made Coffey County
borderline farmers especially ill-suited for survival, much less rehabilitation, in the ever-changing plains farm environment.

Barnes County, North Dakota

The representative county for the plains, Barnes County, North Dakota, is located in the southeastern part of the state, directly west of Fargo. Its seat is Valley City, located in the center of this large county. The Sheyenne River cuts the county into eastern and western halves, and the county's terrain ranges from level to steep, with loamy soils on glacial till plains and moraines, which vary from well to poorly drained. The normal annual precipitation in Barnes County is eighteen inches, making it part of the semiarid Great Plains. Most of this moisture comes in rainfall between May and August. The poor quality of water for human use on many Barnes County farms was a serious issue during the 1930s. Temperatures range from warm summers to frigid winters when Arctic air masses stream over the area. The last frost in the area is normally around the end of May, and freezing temperatures return during the first half of September.[1] Barnes County was organized in 1878, six years after the Northern Pacific Railroad crossed the Sheyenne River. During the 1860s and 1870s European-American pioneers settled the area around Valley City. These immigrants were primarily Irish, Scottish, English, and Canadian in background. The Great Dakota Boom of the 1880s opened the way for German, Norwegian, and Dutch settlers to settle rural Barnes County, and they stamped it with a strikingly varied ethnic cast.

Barnes County was solidly agricultural during the 1930s. Its rural-farm population density, seven people per square mile, was close to that of the plains norm. One distinction between Barnes County and most plains counties was that it saw little change in the total number of farms between the beginning of the Great Depression and the end of World War II. Barnes County farms were also slightly less in value compared to the average plains farm. During the thirties in this often-harsh environment, Barnes County's farms were predominantly dedicated to growing field crops, mostly cash grains such as spring wheat. The remaining farms were livestock, subsistence, and dairy operations.[2] This was Spring Wheat Belt farming, along with considerable livestock growing. Like farmers in other semiarid areas of the plains, Barnes County farmers reeled between the extremes in their income and land values. During periods of low wheat prices, plains farmers normally reacted by increasing acreage in grain to compensate for price drops. However, during the 1930s seven years of rainfall below normal levels kept Barnes County farmers from boosting production of wheat and other crops and livestock. This drove individual farm income in Barnes County below that of the twenties until 1941.

As savings dwindled during the early 1930s, many Barnes County farm families fell into indigence. Farmers Union representatives in Barnes County reported that in the farm towns of Dazey, Kathryn, Litchville, and Pillsbury, families lacked coal, blankets, clothing, and seed wheat in January 1934. These families also needed credit to keep both their farm operations and households afloat.[3] Signs in the early thirties suggested that Barnes County farmers were sinking further into debt and deprivation. Under the Federal Emergency Relief Administration, Barnes County residents received nearly a half million dollars in relief. The federal government bore three-quarters of this cost and local funds covered the rest.[4] Between 1931 and 1934 the federal government extended $336,000 in feed and seed loans to Barnes County farmers. By the end of 1934 three-quarters of this amount remained unpaid. In early 1936, 220 county families received an average monthly grant of $16 to stay afloat.[5]

Borderline farmers in each county made their own particular demands on the rural rehabilitation program. In Barnes County rehabilitation clients required additional grants for subsistence and loans to pay off debts and to diversify from dependency on spring wheat to feed crops and livestock. Also, if possible, Barnes County farmers needed tractor power to cut operating costs for their large acreages. Using a tractor also allowed these North Dakota farmers a higher platform of productivity once the rains and better markets returned.

Coffey County, Kansas

Coffey County, in the east-central part of Kansas, south of Topeka, is only half the size of Barnes County. Burlington, the county seat, is located in its center. The topography of the Coffey County is mostly rolling plains, and the soil is made up of silt loams and silty clay loams. The Kansas Flint Hills ranching region skirts the western part of the county. Annual precipitation in Coffey County fluctuates widely. For example, though the norm is thirty-seven inches of precipitation annually, the area received a phenomenal sixty-five inches in 1941. In an average year, two-thirds of the moisture falls between May and September. The growing season in Coffey County is substantially longer than that of Barnes County; in east-central

Kansas the last frost is generally in the middle of April and the first frost in late October.[6] Coffey County was organized in 1855. In contrast to Barnes County, in the 1930s nearly all farmers were native-born Anglo-Americans, though some were German American.

Coffey County's economy and population were also agricultural. The Kansas county had roughly the same number of farms, and the same proportion of farm tenants with relatives as landlords, as Barnes County. However, Coffey County rural neighborhoods were nearly twice as densely populated, and saw a much greater flight of farmers between 1930 and 1945, than southeast North Dakota. In this wetter, warmer climate, most farmers raised livestock, followed by field crops, then subsistence farming. There were also significant numbers of dairy and poultry farmers.[7] Coffey County farmers grew corn and sorghum to feed their cattle and hogs. Despite climatic advantages, Coffey County had long-term agricultural, economic, and environmental shortcomings. The geographic instability of tenants, low value of farms, comparatively low productivity, low mechanization, less self-sufficiency, and maladjusted farming operations offset its bountiful rainfall and more temperate climate. Rural overpopulation and substantial water erosion gave Coffey County the attributes of rural poverty in border states such as neighboring Missouri.

Coffey County had fewer severe market and climatic fluctuations between the two world wars than did Barnes County. Land prices took a deep plunge after 1920, but not as precipitously as in Barnes County or in the plains in general. As in most of America, the value of Coffey County farm production plummeted between 1929 and 1932. During the thirties the annual rainfall lingered below normal levels in all but three years. Therefore, as in Barnes County, crop failure and low agricultural prices conspired to drive down Coffey County farm income. However, Coffey County farmers made slightly higher incomes than their North Dakota counterparts during the 1930s.

Correspondingly, Coffey County seems to have escaped the chronic poverty levels of many Barnes County farmers during the early 1930s. Under FERA, the Kansas county required (or at least received) only one-quarter of the relief dollars used by Barnes County.[8] Coffey County may have also been better off than its neighbors. For example, in nearby Douglas County, Kansas, there were 130 farm families on relief in 1935 before the federal rural rehabilitation program took over. Coffey County, with a like number of farms, had only 28 farm families on relief.[9]

The farm economy in Coffey County, Kansas, was more stable, but offered less opportunity, than that of Barnes County. In contrast with the rest of the plains, the Kansas county's rural neighborhoods were crowded with undercapitalized, smaller, horse-powered farms. In this region of plentiful rainfall, farms were less dependent on wheat and other commodities with precarious markets. The result was a more gradual drop in income during the 1930s compared to Barnes County.

These conditions colored the goals of Coffey County's rural rehabilitation program. Despite their small farms, Coffey County farmers produced as much income from their operations as other plains farmers did with more land. However, while borderline farmers in Coffey County were more secure than their Barnes County counterparts, they were less able to increase their scale of farming during World War II. In the meantime, the Coffey County rehabilitation program needed to help borderline farmers build up livestock operations, increase income, and cut debts and expenses. Since many of these farmers were on small 140-acre farms, it was not cost-effective to encourage tractor farming. However, the program might have increased long-term productivity by supporting soil conservation practices.

Clientele Portrait

Comparing these rural rehabilitation client farms and their operators with average farms in each county reveals a great deal. Both Barnes and Coffey County rural rehabilitation clients were overwhelmingly renters and were close in age to other tenants in the plains region. But in two categories, acreage and income, the Barnes County clients were at a disadvantage compared to the county norm. As noted earlier, one advantage for plains tenants, at least in theory, was that their lack of high mortgages and property taxes enabled them to afford to rent more land and thus cut farm expenses per acre. Barnes County clients, however, missed this benefit, since their farms were 16 percent smaller than the average farm operation in the county. This led to decreased productivity and an income one-quarter less than the average farmer's. Furthermore, in Barnes County the proportion of rehabilitation clients' farm size to that of all farmers in the county was precisely that of borderline farmers' land to general farmers' land in the 1935–1936 National Resources Planning Board study. This means that Barnes County clients fit the profile of the plains borderline farmer.[10]

For the sake of clarity, I have selected a farm family from each county to illustrate both general farm conditions and the progress of rural rehabilitation efforts in distinctly different situations. All names from case files are pseudonyms. For Barnes County, North Dakota, the farm couple is Lawrence and Anna Schultz; for Coffey County, Kansas, the couple is Ellis and Nora Pound. Neither couple should be considered typical of their respective areas. It is difficult to find a "norm" for farm families, with their varying agricultural, ethnic, social, and economic attributes. However, these two couples offer us an accessible portrait of farm life in their county because their ethnic background, farm layout, acreage, net worth, and gross income closely correspond to those of their fellow rural rehabilitation clients. All the farms covered in this survey were run by men. Rural rehabilitation supervisors took this for granted and focused their reports on the farm husband, often to the exclusion of the farm wife. Therefore, in most cases I must refer to the male client when discussing a farm's overall operation.

Lawrence and Anna Schultz, the Barnes County client couple, had five children: two sons and three daughters. Mr. Schultz was in his mid-fifties on the eve of World War II. His age is important; because of his years of acquired experience, social and economic contacts, and saved capital, the Schultzes were stable renters in the late 1930s. During this time they managed to rent between 320 and 480 acres of land. The Schultzes ran a beef and dairy operation while raising wheat, oats, and millet for sale and corn for silage feed for cattle. Like other FSA client and nonclient farms in the region, the Schultzes used their crops, mainly spring wheat, to make from one-third to one-half of their income. The rest of their income came from cattle and dairy sales and from AAA payments. Until their farm income took off in 1940, the Schultzes grossed between $961 and $1,092 annually in the late 1930s, making them one of thousands of plains borderline farm couples. Lawrence and Anna Schultz's farm operation resembled others in their neighborhood. Out of their total farm income in 1942 they made 58 percent of it from the sale of livestock or livestock products. That percentage is only slightly higher than the average Barnes County farmer's. The couple did depend more on the sale of dairy products than the average farmer in their area. The remaining 42 percent of the Schultzes' income came from the sale of crops, again, close to the average in their county.[11]

The Schultzes were fortunate in having a stable farm tenure arrangement. When they took out their first rural rehabilitation loan they had only a one-year lease without a renewal clause. Still, they were able to remain on the same land between 1938 and 1942. The Schultzes' lease allowed them to keep a two-thirds share of their crops while they paid an $80 annual cash rent for 480 acres.[12] They were lucky to have such favorable terms, since most Barnes County farmers had to split their crops fifty-fifty with their landlord. Under the lease the Schultzes probably had to provide the labor, feed, seed, draft animals, and machinery. Their landlord presumably provided a third of the threshing expenses, and, of course, the land.[13]

In contrast, farmers in Coffey County generally ran operations half the size of the average Barnes County and plains farm. However, Coffey County farmers were able to squeeze income equal to that of the plains average from their smaller acreage because of its greater rainfall, diversification of crops and livestock, and more intensive farming practices. Rural rehabilitation client farmers in Coffey County, Kansas, however, did not fare well in comparison with their fellow clients in North Dakota or with their neighbors in Kansas. Their farms were three-quarters the size of the average operation in Coffey County.

Ellis and Nora Pound, our example for Coffey County, also had five children: three sons and two daughters. And like the average tenant and FSA client, Ellis Pound was in his early forties when he had to take out his first rural rehabilitation loan. The Pounds' farm acreage rose and fell until it stabilized at 120 acres beginning in 1940. It was more committed to livestock raising than most farms in the area. In Coffey County, farmers made up to three-quarters of their income from livestock and livestock products and the remainder from the sale of crops.[14] The Pound farm depended almost totally on raising livestock on its small acreage, with only 6 percent of its income coming from field crops, and that was from a commodity payment. The couple attempted to combine a cattle-raising operation with some dairying. About seventy acres of their land was in pasture, and the remainder was in corn, sorghum, and oats which they used to feed their livestock, as was the standard practice in the county.

The most striking aspect of the Pounds' farm was their complete lack of secure tenure. Between 1939 and 1942 the couple moved every year. Each rented farm required one-third of their crop for rent, as well as an annual cash rent ranging from $75 to $200. In 1942 the Pounds moved to a farm with a three-year lease that may have allowed them to sink some roots—for a while at least.[15] Ellis and Nora Pound's example shows the difficulties of lower-tier tenancy in Coffey County. Conceivably renters enjoyed opportunities to shop around for better lands and better leasing arrangements, but such extreme geographic mobility must have caused hardships. Moving so often increased tenants' expenses, lessened familiarity with their soil, frustrated long-term planning, and damaged their local reputation among prospective landlords. The Pounds' inability to stay on a farm more than one year may have lessened their economic stability as well. Since they kept their most valuable assets—livestock—their sole income during these years was from the sale of dairy and eggs, rural rehabilitation grants, and work off the farm, including work with the WPA, until their income took off in the early 1940s. The Schultzes' relative stability and the Pounds' lack of stability are indicative of conditions in each county.

The clients' case files reveal even more about their finances. Client farm families in both counties had the same amount in debts. Yet Barnes County clients had 20 percent more value in the assets of their farm and nearly double the net worth of their Kansas counterparts. In addition, Barnes County clients faced 60 percent higher farm and home expenses as well as farm receipts projected by the FSA for the 1941 crop year.

Comparing the Schultz and Pound farms in 1940 reveals that the total value of the Schultz operation was fairly high at $3,200, which was unchanged from the previous two years. Most of the value of the operation was tied up in livestock, particularly beef and dairy cattle, as well as draft horses. Much of the Schultzes' feed and fodder for the animals came from the farm. When they became FSA clients, most of their debts originated from year-to-year expenses for additional feed and seed, livestock, small implements, and past-due cash rent, totaling a disturbing $2,700 in 1936. Up until 1939 the Schultzes' total debt for rural rehabilitation loans was $140, which grew to $1,400 by 1940. They had repaid their debts to various creditors with Mr. Schultz's FSA loans. The Schultzes' larger-scale farming operation meant heftier expenses for running the farm and farm home, and FSA funds allowed the family to greatly increase their spending. Cash farm expenses grew from $250 in 1936 to $1,200 in 1940. During the same time their cash spending on the home increased from $280 to

$680. These high expenses left the Schultz family with virtually no net income in the late 1930s, and in 1940 they were actually in the red for $86.

FSA farm supervisors were constantly wrong about their clients' estimated future income. For most years the income projected by the FSA office was off the mark by one-third from the actual income. The farm supervisor had unrealistically high expectations for the Schultzes' farm income until 1940, when actual and projected incomes were closer to each other. Worse, the FSA's hopes for *net* cash income after subtracting expenses were totally impractical. While the projected income from the sale of crops and projected expenses were close to actual, the goals for increased sales of livestock and livestock products were unfulfilled.[16]

In many ways, Ellis and Nora Pound's farm operation was similar to the Schultzes'. Most of the Pounds' assets were in the form of livestock, feed, hay and fodder. One-third of the value of their livestock was in their draft animals. The Kansas couple's assets totaled $1,600 in 1940, close to the Coffey County average for rural rehabilitation clients. The value of the Pounds' assets was fairly constant between 1937 and 1940. The following year their assets rocketed upward. Between 1937 and 1942 the Pounds' liabilities of $1,300 were equally stable. Most of their debt was for livestock purchases. Also, during this time the couple had a running bill with a local grocery store for $40, as well as other small debts. Although these were not crippling obligations, their inability to remit them reveals their low income.

To meet these debts, the Pounds applied for and received a rural rehabilitation loan. Their debt to the FSA grew from $400 in 1936 to more than $1,000, where it remained through 1942. Although the Coffey County family was able to settle their livestock and other debts, Ellis and Nora Pound made slow progress in repaying them. Their net worth was unusually low in the late 1930s; in 1940 it remained at $258, but within the next two years it increased greatly.

Like the income projections for the Schultzes in Barnes County, those for the Pounds in Coffey County fell short of the target in 1940. However, perhaps by scaling back on expectations, the FSA's income projection for the Pounds' income that year was close to their actual $1,122 in farm receipts. In the late 1930s the FSA's income projections for the Pounds' crops were altogether overly optimistic. By 1940 the FSA dropped crop sales from the projections completely. Like the Schultz farm, the Pound operation had no net cash surplus until 1941, and then they were $85 in the black for the next two years.[17] What was especially distressing about the Pounds and other clients in the Kansas county was their small annual income of $550. Coffey County clients grossed less than half the farm receipts of the average farmer in their area, and one-third that of Barnes County clients. Therefore, they must have stood out in their neighborhoods to a greater degree than rural rehabilitation clients in Barnes County. Put another way, while Barnes County clients were in the upper ranks of borderline farmers, their Kansas counterparts were at the bottom.

Taken collectively, the data on Barnes County and Coffey County rural rehabilitation clients gives us a rough portrait of their farming operations in 1940. In both counties clients' operations were comparatively small-scale, both in acreage and in capital investment. In both counties the clients were overwhelmingly tenant farmers, usually in their early forties. Finally, they were part of the group of borderline farmers grossing roughly between $500 and $1,000 annually. However, there were differences between the clients in these two counties. In Coffey County they were in the lower ranks of borderline farmers. Their incomes placed them on the edge of poverty and they had diminished chances for raising their income. True, from a financial perspective, Barnes County client farmers had their weaknesses. Fewer of them owned their land, they had higher farm and home expenses, and had acquired more FSA credit to pay for them. Most troubling was that, despite their high capital costs and large acreages, these Barnes County farmers were still skirting the poverty level.

Nevertheless, compared to their Coffey County counterparts, Barnes County clients were able to spread their liabilities and expenses over more acres. This is particularly so for Lawrence and Anna Schultz, who held $3.32 in debt per acre compared with Ellis and Nora Pound, who held treble that amount—$11.18—in debt per acre. The Pounds probably added more debt to increase the scale of their farming operations. But unlike the Schultzes, they started the rehabilitation process with insufficient acreage and a low value of assets in livestock, machinery, feed, hay and fodder, as well as household goods. The Pounds, like other Coffey County farmers, started off with higher debts per acre and fewer assets compared to those in Barnes County. Coffey County rehabilitation clients also had comparatively smaller farms, greater geographic mobility, and lower income. This was the stark situation facing the rehabilitation process as it began in Coffey County.

Fixing the Farmer: Reasons for Obtaining a Loan

How did the Resettlement Administration and the Farm Security Administration attempt to transform the farming operations of our two representative rural rehabilitation clients? To fully understand these attempts,, we need to look at the reasons for the loans and the overall progress in rehabilitation, that is, progress in income, debt reduction, and better tenure arrangements. Clients' reasons for taking part in the rural rehabilitation program and their progress in it differed by county. Because of this, certain conditions made rehabilitation more viable in some locales than in others.

Farmers in Barnes County, North Dakota, received rural rehabilitation loans to reduce their debts, recover from crop failure, and change the scale and orientation of the farm. Many farmers wallowed in their financial obligations and couldn't go to banks for needed credit. For example, a local bank pressured rehabilitation client Spencer Ames to pay off $5,700 in loans in 1939. Clearly he was unable to satisfy the bank because of his low income and insufficient security. Ames also owed a total of $200 to local agricultural implement dealers on debts for a tractor and for a cream separator.[18]

The Barnes County farm supervisor, Alden Baillie, noted that if client farmers kept up with their repayment schedules on their debts, it would hurt chances for future rehabilitation. One example of this was tenant farmer Fred Dietz, who lived in a part of Barnes County rich in hay and pastureland. In early 1939 the Reconstruction Finance Corporation pressured Dietz to repay his balance of $500. To do so, Dietz would have to sell off his livestock, which would cripple his operation, according to Baillie. "If he is enabled to keep his livestock for two or three years he will be in splendid condition to pay off a much larger debt," Baillie claimed. "No amount of supervision will be required as he is a proven operator and at least an average manager."[19]

Another reason Barnes County farmers applied for a rural rehabilitation loan was indigence caused by drought conditions in 1938–39, when area rainfall was only thirteen inches annually for the two years. Drought affected individual clients differently, so rehabilitation needs varied by each farm. For example, Albert Bemmer had suffered an unfortunate string of crop failures before the drought struck. Despite the comparatively ample rainfall in 1937, his wheat crop yielded only one bushel per acre, while he harvested passable yields in oats and corn. During the following years of parched fields and pastures, Bemmer had fed all his crops to his livestock. Under normal conditions, wrote farm supervisor Walter Stine, Bemmer should have been able to feed his cattle, sell some of his produce, and pay his debts from his farm receipts. But because of depressed conditions, without additional FSA credit Bemmer might have to sell off his herd. Therefore, Stine recommended that the farmer take a loan and build up his breeding stock for the long haul.[20]

Another example was Ernest Jones, who in early 1939 was already $1,355 in debt to the FSA. Drought, grasshoppers, and hail hit this father of nine children hard on his 480-acre farm. The Joneses' family size seemed neither to hurt nor help their finances. In 1940 the couple had only one son, aged fifteen, who probably labored in the field. On the other hand, three of the Joneses' daughters, ranging in age from sixteen to twenty, probably helped produce the $120 in dairy sales for that year. The home supervisor noted that the tenants lived in a sparingly furnished, drafty, seven-room house and spent $300 annually on home operating expenses—not much more than the Schultzes spent for their moderate-sized family. Walter Stine wrote that although the Joneses and other client families hadn't made progress, "they have been holding their own and making a satisfactory living for their families." Stine recommended a supplementary loan to cover the $1,000 in projected farm and home expenses for the following year.[21]

Barnes County client farmers also took out a rural rehabilitation loan to increase the scale of their operations or at least profit more from them. Client Richard Chisholm explained that if he hired his 260 acres of hayland cut, the cost would wipe out the amount of hay needed for his herd. He therefore requested a loan of $100 to pay for repairs on his harvesting equipment and pay for labor.[22] Supervisor Stine showed some discretion in submitting loans for clients to mechanize their farms and cultivate more land. In spring 1941 Stine recommended a loan for Sidney Compton to purchase a tractor for his 320-acre tenant farm. Compton, in Stine's opinion, had already advanced his operation's status by building up his herd to twenty cattle and by securing a sound lease on his current property. Only the disrepair of the client's old tractor had held him back, according to Stine.[23]

Finally, Barnes County clients requested rural rehabilitation loans to change their current farming operations. As noted, the RA and FSA pressured the region's farmers to transform into small-scale, semisubsistent livestock enterprises. While making the transition, clients continued to repay past debts, support their families, and pay for capital investments. For example, when Cletus Fischer applied for a rural rehabilitation loan, nearly all his cropland was in wheat and oats. The required conversion from a cash-grain operation to one envisioned by the RA and FSA was no easy task. While Fischer was switching some of this property over to pasture, he needed credit to pay off the balance of a loan for a tractor and a previous mortgage, and to purchase six additional milking cows.[24]

John Bollinger was an exceptional client who owned a 960-acre, fully mechanized farm with one-third of its cropland in wheat. Bollinger was in dire straits, as he owed $11,500 in arrears: $10,000 in mortgages, $880 in obligations for agricultural equipment and an automobile, and the rest in miscellaneous debts. Normally such a substantial farmer would not have been able to float a loan with the Farm Security Administration. However, the FSA supervisor saw the farmer's predicament as a good opportunity to switch from his current "cash grain program," to a livestock program complete with cattle, sheep, and hogs.[25]

Rural rehabilitation clients in Coffey County, Kansas, also needed credit, but for sometimes different reasons. Coffey County clients received loans for relief from crop failures, to increase the size of their operations, improve farm management, get a start in farming, and address soil erosion. Borderline farmers in the area looked to the Resettlement Administration for relief from the crop failures of 1936. With twenty-three inches in precipitation that year, east-central Kansas had a comparatively severe drought that withered corn and sorghum crops and pastures used to feed livestock herds. Because of the low rainfall in 1936, Coffey County farmers fought to keep their operations afloat. H. A. Dressler, the county's assistant farm supervisor, noted that one stockman, Erwin Domer, was "one of the best farmers in his community." Domer owned his fertile 160-acre farm, which he cultivated with a tractor. However, in 1936 chinch bugs destroyed forty acres of his wheat, and drought and grasshoppers ravaged his remaining crops. Local banks presumably denied Domer a loan since he already owed $4,000, or 80 percent of the property's value, on a mortgage to the Federal Land Bank, and $100 in past due interest. Dressler recommended an $85 loan to reseed Domer's pasture and plant crops for the next year.[26]

Many small farmers in the region sought to increase the scale of their operations. The pressure to find and enlarge one's landholding must have been extreme. The high annual turnover of tenants on Coffey County farms both exemplified and exacerbated the problems. Rehabilitation clients like Ellis Pound found it tremendously difficult to secure and maintain larger farms to keep up with the increasing scale of farming during the late 1930s.

Elmo Botsford's experience also showed the difficulty of advancing the scale of farming. In early 1938 Botsford applied for a $380 rural rehabilitation loan. He had recently doubled his acreage by moving to a 160-acre tenant farm. The farmer now had the potential of advancing from modest-scale agriculture to a larger operation where he could increase his productivity and his income. Perhaps he could advance to an even larger tenant farm—or even become a farm owner. In the meantime, Botsford required the FSA loan for additional equipment and dairy cows. Unfortunately, two years later Botsford had moved again to another tenant farm—and in the process shrunk back to eighty acres.[27] As with so many other Coffey County rehabilitation clients, he was unable to preserve his business, much less expand to the larger acreage to keep pace with the agricultural demands of his time.

Coffey County had unique problems that contributed to the shortfalls among rural rehabilitation clientele. FSA case files from the county often blamed poor managerial ability for low income and inadequate farming operations. Whether the supervisor in this county was especially critical of his clients or whether smaller farms attracted less-talented farmers is unknown. The much-afflicted 160-acre farm of Herman and Lois Eddy illustrates how much managerial ability it took to run a commercial farm during the Great Depression.

The Eddy farm was smaller and more diversified than many Coffey County farms. The Eddys had twelve plots of land within their farm, in addition to the farmstead. In planning out the year's operation, the couple had to use all of their forty years of experience in agriculture to successfully manage their farm. They had to decide whether land remained in pasture or in crops. They also had to plan the amount and disposition of their crops by taking into account changing factors such as government policy, commodity markets, soil fertility, rainfall, and the cost of labor and farm machinery to work the land.

Most important, the Eddys were lower-tier renters. In 1935 the couple had been on the same property for only two years. Knowing that they might have to move to another farm the following year (which they did), they had to reap what they could without investing too much in property they would leave behind. Long-term FSA goals of building a stable, profit-able, self-sufficing operation conflicted with the short-term realities of ten-ant life. Since the Eddys were renters, two factors stymied any advanced plans. First, their landlord could rule out growing soil-building crops if he felt that doing so might reduce his share. Second, their mobility undercut the Eddys' plans for building up the ideal farm operation.[28]

Although our focus is on the agricultural and economic factors in-fluencing the lives of plains farmers during the 1930s, it is difficult to sep-arate these factors from the effects that social and family relationships had on the progress of client farmers. Therefore, it would be inaccurate to look at the extreme mobility of Coffey County farm tenants from a solely agri-cultural or economic perspective. Personal and social factors had strong implications for the lives of borderline farm families.

Despite pastoral images, farming could be extremely stressful. Arbitrary climate, unsympathetic creditors, and changing agricultural conditions alone took their toll on mind and body. To a remarkable degree, decisions on Herman and Lois Eddy's farm must have been based on noneconomic factors, such as the health problems that afflicted the couple and their three children. The home supervisor reported that Lois had high blood pressure, which she tried to control by diet. Because of her poor health, Lois couldn't perform much housework in the summer heat. The family's options were further limited by the children's poor health. The Eddy's youngest daughter, Bernice, in her mid-twenties, was "mentally inef-ficient" (mentally handicapped?), in the home supervisor's words. The su-pervisor reported that scarlet fever had left Bernice with one limb shorter than the other and made her self-conscious as she performed the house-work, cooking, gardening, and canning. Their son Les, in his upper teens, was "mentally incapable of working alone," and the family could not afford a truss for his hernia. Unfortunately, the Eddys were unable to pay for vis-its to a doctor. Despite their problems, the Eddys still received high marks from the home supervisor. On a visit in 1936, Lois showed "initiative, resourcefulness, and managerial ability" in running the home. "The family is trying to make the best of things at hand," the supervisor reported. "I believe they will eventually make good."[29]

Under such conditions, keeping the family together and making the farm profitable would have taxed the best managerial mind. Consequently, the home supervisor was perhaps optimistic in justifying further aid to families like the Eddys. Coffey County rehabilitation clients had to walk a fine line because they had fewer resources—and less room for error—than did Barnes County clients. For these reasons, Coffey County client farmers needed advice on operating their businesses. A striking number of Coffey County clients required, in the farm supervisors' judgment, management assistance in running their farm.

Rudy Docker, whose seven years on the rural rehabilitation program rolls were the longest continuous stay of any of the clients surveyed in the Coffey or Barnes County files, is one example. During this time, between 1936 and 1942, Docker worked his father's 115-acre farm. In late 1939 farm supervisor Dressler reported that chinch bugs and drought destroyed Docker's crops. Natural conditions that would have bedeviled a skilled farmer drove him out of business. Although Docker was considered honest, remarked the farm supervisor, "his reputation for good hard work is not the best in this community." Because of this, Docker was "badly in need of supervision" in attempts to diversify his crops and convert to a livestock and poultry farm.[30]

In addition to problems caused by poor managerial skills, the Coffey County rural rehabilitation program had to deal with a fair number of farmhands entering farm tenancy. Young couples could afford tenant farming in Coffey County, where the average value of a farm was only three-quarters that of one in Barnes County. Warren Boldt is an example of a client attempting to make the jump from farmhand to renter on a tiny forty-acre property. He leased the land from his uncle for the sweetheart price of the payment of taxes and upkeep. Boldt applied for a small FSA loan in late 1937, when he ran the farm with only one cow, agricultural machinery valued at $31, and the use of his uncle's tools. The total value of the Boldts' possessions was a scant $311. Despite his small-scale operation, Warren Boldt was, according to the farm supervisor, both "industrious [and] reliable," and his farm had a promising future. Boldt made $350 that year working off the farm, had no outstanding debts, and was willing to barter with neighbors for machinery and additional pasture. He requested a rural rehabilitation loan to purchase a couple each of horses, dairy cows, and hogs.[31] Other farmhand families applying for rural rehabilitation loans had similar attributes. They were young people, with inadequate income and property but few debts, who wanted in the farming game.[32]

As a final goal of their farm operation, Coffey County clients used loans to correct soil erosion. Barnes County, North Dakota, undoubtedly had its problems with depleted land, yet this was never mentioned in its surveyed case files, probably because of the area's farming practices and rehabilitation strategies. Barnes County farms were large and often devoted to cash crops. The county's rural rehabilitation program might have seen diversification away from spring wheat as benefiting both farmers and the land. Therefore, a separate soil-building program may have seemed superfluous. On the other hand, Coffey County farms were limited to smaller acreages and benefited most from enhancing the land's fertility. In addition, some areas of Coffey County had suffered substantial soil erosion because of poor drainage during the 1930s and needed restoration.[33] It appears that much of the silty loam soil in the southeastern Kansas county was better suited to grazing than to cultivation. In addition, the western edges of the county were part of the Flint Hills ranching area.[34] Apparently many smaller Coffey County borderline farmers were attempting to survive on comparatively cheap livestock operations that were too small to succeed as ranches. Therefore, FSA clients in the area often lacked the acreage and good soil to make them profitable as livestock and grain farms.

Attempts to restore the land in Coffey County provide an illustration of the soil conditions in the eastern plains. Although the farm supervisor lauded client Art Claire's soil terracing, the farmer also seemed to be moving away from mixed farming toward cash crops. In 1938 he had his land divided between cash crops and soil-restorative feed crops. However, by 1941 Claire had converted to half wheat and half feed crops, perhaps in response to wartime markets.[35] Supervisor Dressler encouraged clients to plant lespedeza, a crop used to revegetate poor pastureland, and client William Coffer's landlord provided him the seed to build up the pasture.[36] Not all landlords were amenable to this. Victor Dunbar, another Coffey County client, was eager to take part in the second AAA's soil conservation program. Under this plan, he would have taken his land out of cash grains such as corn and plant soil-conserving crops such as alfalfa. Dunbar rented additional land to build up his soil-restoring operations; however, his landlord insisted that the land be planted in corn and that he receive three-fifths of the harvest.[37]

In addition to crop failure, debts, and low-scale productivity, Coffey County farms faced further problems. Compared to their North Dakota counterparts, Coffey County rehabilitation clients seemed to have more management difficulties. With their small acreages and inadequate resources, farm couples such as Herman and Lois Eddy had a smaller margin of error than their larger neighbors. Consequently, they were more vulnerable to poor health and more in need of nonfarm income. Also, Coffey County had more families just making their entrance into farming. Although these were young couples full of energy, drive, and skills, their lack of essentials—land, equipment, and credit—made rehabilitation an uphill battle. Finally, soil erosion was a serious problem. Whereas many clients attempted to practice soil conservation, they were also under substantial pressure to devote themselves to cash-grain crops. Between landlords and rising wartime commodities markets, these clients, who were mostly tenants, were sorely tempted to plant as much corn and wheat as possible to increase their income. Of all the above issues, the most important was boosting the scale of farming. In Coffey County the presence of many novice farmers with few resources or limited managerial abilities must have diminished the rehabilitation prospects for the county's clientele as a whole.

Fixing the Family Farm: Progress

The rural rehabilitation program attempted to improve agricultural income, the domestic and economic performance of farm women, and farm tenure conditions. As noted earlier, state and federal agricultural officials sought varying farming methods for plains farmers pummeled by crop failure and low income. The RA and FSA embraced two solutions: near-subsistence agriculture and diversified farming. This made sense, since these were comparatively inexpensive methods. Also, these methods dealt with the immediate emergency situation: keeping families fed, solvent, and working on the farm rather than swelling relief rolls in nearby towns.

However, the most viable solution was increasing the scale of agriculture. In the plains, this led to a commitment to increased acreage, productivity, mechanization, and outside credit. For borderline farmers struggling to keep pace with the growing scale of agriculture, this meant cutting back on operational expenses for the farm and home, increasing nonfarm income, becoming more efficient, and boosting productivity.

Yet borderline farmers who were rural rehabilitation clients in the late thirties and early forties found it difficult to get a leg up by these means. Clients in Coffey and Barnes Counties were spending only between $500 and $900 annually to run the farm as well as clothe, feed, and house the family. Expenses were already cut to the bone. Sizable nonfarm income was unlikely for clients for several reasons: running a diversified farm was a full-time job, work relief and private employment were scarce and often far away, and the county rural rehabilitation committee rejected applicants who managed to make a livelihood off the farm.

This left increasing efficiency and boosting productivity as the means to keep up with large-scale agriculture. Farm supervisors worked with the client families to make their operations more efficient and productive— but only on a diminished level. As noted, clients used most rural rehabilitation loans to pay off debts, supplement their small livestock herds, repair the farm infrastructure, and purchase some farm and home implements. However, this was not enough. One alternative was a radical reform of farm practices to balance the resources of borderline farmers with their financial desires. This would have required a pronounced readjustment of rural expectations for a standard of living along urban lines. Although the 1930s saw a temporary curtailment of expectations, the destructive impact of drought and depression only served to temporarily slow the drive for greater production and income in the plains. Borderline farm clients were out of their depth. FSA clients in Barnes and Coffey Counties were only able to receive $800 in credit—not enough to make the jump into large-scale agriculture.

Barnes County clients had an average annual gross income of around $900 in the late 1940s, and receipts from their business reached only $1,200 in 1941. At this point apparently most surveyed rural rehabilitation clients had left the program. Coffey County clients showed lower income during the same period. They were stalled at a much reduced annual income level of about $500 until 1941, when their annual income took off and reached $900. Many Coffey County rehabilitation clients stayed on until 1942, when their annual gross income reached $1,250.

The duration of Barnes and Coffey County clients within the rehabilitation program reflected the progress they made. Barnes County clients tended to stay in the program for shorter periods, presumably because they had achieved an adequate income and some control over their debts. Many Coffey County clients, in contrast, still required aid into 1942, when much of the countryside had returned to prosperity.

The economic downturn of the 1930s dealt both farm men and farm women a cruel blow. When the farm economy collapsed during the early part of the decade, the responsibility often fell on farm women to maintain the family's living standards. To compensate for drops in income that would normally come from sales of crops and livestock, women attempted to expand home production and self-sufficiency. Farm women throughout the plains searched for ways to earn money through cottage industries, selling dairy products and poultry, and letting vacant rooms out to boarders. Some even cut their expenses by making their own consumer products such as soap, toothpaste, and hand lotion.

Most government officials, businessmen, and reporters connected to agriculture saw women's work, concerned mainly with small-scale productive chores and housework, as only supportive or an auxiliary of farm men, who produced for the market. However, in the years leading up to World War I, agricultural schools and state extension services took notice of the status of farm women. Academicians and home demonstration agents assessed the farm woman's primary problems as poor home management, overwork, cultural isolation, and low income. As a solution, they advocated running the farm home on business principles. The FSA home supervisors as well saw proper home management as the way to resolve many of the borderline farm women's problems. Within the FSA program farm women had an important role providing economic security for their families, and as a consequence the rural rehabilitation program showed a great interest in them. In addition to reforming work and planning in the field, the rural rehabilitation program also tried to rehabilitate farm women and their work practices and living standards within the farm home. By no means did the staffers wish to change the conventional domestic role of farm women. In fact, the agencies reinforced these roles. The rural rehabilitation program of the 1930s and early 1940s was based on the family farm unit. Its guidelines and regulations assumed a partnership where the husband worked in the field and farmyard, and the wife kept house while gardening, maintaining small-scale dairy operations, and raising poultry.

An important strategy in the rural rehabilitation plan was to convert clients' farms and farm homes to semisubsistence. This meant partially withdrawing both male and female labor from their dependence on the grain market economy. In these small-scale operations, both men and women were to work out their "program" until they were on their feet again. By strengthening the woman's role in the farm operation, this plan could conceivably strengthen her voice in running it. But the RA and FSA did not operate without gender bias. The vast proportion of the loan money went for "male" operations in growing feed crops and raising livestock. In addition, the survey of clients in the North Dakota and Kansas counties found no unmarried women receiving FSA credit. There were, however, some single males living with female relatives who were able to float a loan.

Farm women in both Barnes and Coffey Counties had three goals, according to the local home supervisors. They were to create a modest income, cut domestic expenses, and uphold and increase their standard of living. Anna Schultz's experience probably was typical of Barnes County farm women within the rural rehabilitation process. In the late 1930s the Schultzes were grossing around $1,000 annually—somewhat more than the average rehabilitation client but still indicative of other clients in the region. During this time the Schultz family depended on the sale of crops, livestock, and dairy products for their income. Anna Schultz also sold poultry and eggs, but this was a much smaller source of income, and was shrinking from year to year.[38] In other families, checks from cream sales covered a great deal if not all of the family's living expenses. The FSA was keen to capitalize on this and made loans to clients to purchase cream separators to make their dairy operations more efficient.

Because of the smaller scale of agriculture in Coffey County, the farm woman's contribution was relatively more important to the family income than in Barnes County. During the period with income data, from 1937 to 1942, Coffey County rural rehabilitation clients were below the poverty level in income. Farm women such as Nora Pound kept the family fed through her milk and egg enterprise—along with government aid—during the depression. The Pounds, who appear to have been a bit more successful earlier on than most Coffey County clients, still used cream and egg money to make up half their annual income as the family climbed out of poverty. By 1941, however, the farm couple appears to have dropped out of the cream and egg business completely to concentrate on raising cattle with a few hogs.[39] The experiences of other Coffey County farm women also suggest that cream and egg money brought their families up to the subsistence level. The home supervisor reported that money from cream and egg sales, averaging $10 each week, covered the families' subsistence needs.[40]

However, as important as subsistence production was for Coffey County, it fell short of that of Barnes County. With the Kansas county's wetter, more temperate climate, the farmers there might be expected to have been more self-sufficient, to have produced more food and fuel for themselves. The experience of the beleaguered Eddy family offers a glimpse of self-support in Coffey County. Despite their health problems, Lois Eddy and her daughter raised their own beef, pork, vegetables, and fruits such as berries, peaches, and plums.[41] Clients in frigid, semiarid North Dakota would have been sorely taxed to harvest such a variety of products for themselves. However, Barnes County farm families produced nearly twice the value of home-used products as their counterparts did in Coffey County. Despite the limitations of the climate, families in the North Dakota county with their greater resources and larger farms were able to produce a greater value of food, fuel, and housing for subsistence. One reason for this may have been the concentration of foreign-born farmers and farmers with foreign-born parentage in Barnes County. As noted earlier, plains farmers with foreign-born backgrounds tended to produce a greater value of subsistence goods on their land than did native-born ones.

Farm women's subsistence efforts were important within both the rural rehabilitation program and the depressed economy of the 1930s. As the agricultural markets failed these client families, it made sense to withdraw from them as much as possible until prosperity returned. To a degree, during the depths of the Great Depression there was a reversal of trends greatly affecting rural America. Before the 1930s plains farm children were leaving the countryside, the number of the smallest farms decreased, and the importance of farm women's production in total farm income declined. Yet after the bottom fell out of the farm economy, prodigal youths returned to their parents' land, many moved to small farms to ride out the depression, and farm women's productivity increased in importance in the farm budget. However, once farm markets returned in the early 1940s, many deserted the farmstead as a refuge for subsistence. With the return of economic opportunity on and off the farm, it made more sense to in-

crease production of a few commodities, abandon small-scale dairy, poultry, and garden projects, and buy necessities from a store. Both the Schultz
family of Barnes County and the Pound family of Coffey County did this.

Farm supervisors also attempted to improve leasing terms for their
clients by promoting longer leases. Furthermore, supervisors encouraged
share rents over cash rents and compensation for farm improvements
made by the tenant. However, Barnes County rehabilitation clients who
were renters were fortunate to find land under any conditions. In fall 1940
USDA agricultural economist Elco Greenshields visited Barnes County to
study how to improve tenure conditions for rural rehabilitation clients. As
the countryside was slowly emerging from the economic depression of the
previous decade, Greenshields found larger farmers absorbing small tenant farms. The former were more substantial mechanized operators who
were able to lease and purchase additional lands "in order to spread the
higher cost of machinery," he noted. Half the rural rehabilitation clients in
the North Dakota county still used draft animals rather than tractor power,
and Greenshields found those still using horsepower disadvantaged in se-
curing farmland. The result in Barnes County was a number of empty
farm homes and formerly rented properties combined into larger farms.[42]
Turning south, there is little information on the success of FSA attempts to
improve Coffey County tenancy conditions. As Ellis Pound and other
clients in southeast Kansas must have known, the large number of small-
scale renters seeking a fixed amount of land created a seller's market for
landlords. The results were short leases, high rental fees, and little chance
to improve soil conservation or rental terms. Overall, the rural rehabilitation program failed to alter tenant practices significantly in either county.

Assessing Rural Rehabilitation at the County Level

The Barnes County and Coffey County rural rehabilitation programs present both similar and distinct portraits. On one hand, clients in both counties attempted to build up their assets while cutting their high debts. They
used their precious assets and FSA credit to steadily pay off their loans in
the late 1930s. Clients in the North Dakota county concentrated on cutting
their debts rather than building up their net worth. It is striking, however,
that from 1936 to 1941 clients in both counties had almost exactly the
same amount of debt at the beginning—$1,500 in Barnes County and
$1,300 in Coffey County—as they did at the end of the program. Also, be-

tween 1938 and 1942 the total average debt to the FSA remained stable, between $700 and $800. Clients were substituting old debt to the usual creditors with the less pressing rural rehabilitation loans. Therefore, clients from both counties were "farming in place," compared to the more successful plains farmers who were expanding their operations and profiting from the improved commodities markets during World War II. Both Barnes and Coffey County rural rehabilitation clients failed to keep pace with the scale of plains farming, which was increasing even during the economic doldrums of the late 1930s.

However, to be fair to the rural rehabilitation program, elevating and transforming farm operations was expensive and difficult work. This was particularly true under depressed farming conditions. Changing the course of plains agriculture was like trying to drive a moving tractor without a steering wheel. Market, agricultural, technical, and financial forces combined to accelerate the trend toward larger-scale farms in the region, and the FSA lacked the resources to equip clients to meet these trends.

Rural rehabilitation attempted to correct some of the long-term problems resulting from established agricultural practices and conditions in each county. In semiarid Barnes County, the program encountered a legacy of erratic income from cash-grain farming. Hampered by the drought and low commodity prices of the 1930s, the comparatively large farmers of Barnes County had diminished incomes and high debts. The farm supervisor's answer was to convert from cash grains to livestock and forage-crop farms. In Coffey County, Kansas, the program met agricultural practices more in line with its program. Farms there were already devoted to raising livestock and forage crops. However, client farmers in the Kansas county had substantially lower incomes, lower farm values, and smaller acreages. Considering the liabilities and costs on their smaller tracts, Coffey County clients had higher debts and expenses per acre compared to those in Barnes County. FSA supervisors encountered the same problems of high debts, inadequate acreage, and crop failure in both counties. But in east-central Kansas, the FSA also had to contend with soil erosion, inexperienced farmers, and poorly capitalized farm workers as they attempted to run successful tenant farms.

Was rural rehabilitation in each county a success? Concerning the elementary goal of keeping farm families in both counties fed, housed, clothed, and their farms running, the answer is an unqualified "yes." After years of low income, living conditions on plains borderline farms were deteriorating before the rural rehabilitation program lent assistance. Within the case files of both counties, staff described poor diets and housing and farm buildings sadly in need of repair. FSA grants and loans gave hope and a second chance to families living in difficult circumstances. Adverse economic and environmental conditions pummeled tens of thousands of plains rehabilitation client families. Urban-oriented relief programs neglected them. Therefore, for these families the rural rehabilitation program was a godsend.

On a secondary level, assessing the FSA's track record in the two plains counties is more complex. Most supervisors expected clients in each county to go beyond sustaining their families and maintaining their farms to increase their productivity and position themselves to benefit from future prosperity. Evidence limits this study to tracing families' progress while they were rural rehabilitation clients. After that point we can only speculate on their future using census data. To that end an important measure is the total number of farms in 1930 and 1945.

By this criteria, Barnes County clients were more successful than those in Coffey County. A much greater proportion of farmers remained in the North Dakota county than in the Kansas one. Furthermore, Barnes County borderline farmers were more productive before Pearl Harbor, and by 1945 had increased their productivity to a greater degree than in Coffey County. Coffey County farmers were especially vulnerable in the shifting farm economy. Looking through the rural rehabilitation case files for the Kansas county, one runs across several clients selling out during World War II while the rest of the countryside enjoyed relative prosperity. Coffey County home supervisors reported simple "bad luck" followed the lives of these client families and degraded their living conditions. It seems a black cloud followed couples like Avis and Betty Corbin and their three children. Badgered by crop failures and health problems, the FSA advanced them $1,000 in loans. In 1941 the Corbin home burned down and the couple had to move to the barn loft, and then into a neighboring town until the company that owned the land provided a new home. In early 1943, with $1,800 in debt owed to the government, the couple sold their livestock, machinery, and other possessions. Presumably they left farming entirely.[43]

The trials of Coffey County couples such as the Corbins or the Eddys lead us to ask whether they belonged in the rural rehabilitation program.

Although the agricultural hazards of the 1930s tested the best of farmers, these "black cloud" farmers may have lacked the competence to manage a complicated business enterprise effectively. It made sense to sponsor these families during the Great Depression, since there was little opportunity elsewhere. However, after 1941, when agricultural income soared, assisting these farmers was less tenable. Some like John Black proposed a separate rural rehabilitation program for lower-tier "problem" clients like the Corbins and the Eddys. This group, containing many of the Coffey County clients, needed longer-term assistance, guidance, and perhaps a program to ease them out of the countryside and into war industries. The upper tier of the rural rehabilitation program, consisting of more promising borderline farmers, more typical of the Barnes County clients, remained on the farm to become middle-income farmers.

For political reasons the Farm Security Administration sustained distressed families such as the Corbins on their farms only while the country endured widespread economic depression. When prosperity returned again in the early forties and the FSA lost its funding, it was every family for itself. This was not inevitable, however. Despite the return of prosperity, the federal government could have continued its constructive role in assisting the thousands of borderline farm families across rural America. If not for the goal of promoting a vibrant rural culture, the rural rehabilitation program could have continued to acclimate smaller farm families buckling under the demands of modern large-scale agriculture. However, to the powerful farm and business organizations such as the Farm Bureau and the Chamber of Commerce, which advocated larger farm operations, the rural rehabilitation program's regimen of diversification and self-sufficiency must have appeared quaint during World War II. Worse, for those promoting full agricultural production for the war effort, such a program appeared obsolete and an obstruction to prosperity and the war effort.

This two-county study of the plains FSA program shows that it faced a continual struggle between its long-term planning and the short-term needs and desires of its clients. Some of the trends of the times fit with the FSA's designs for "secure farming," and some ran counter to them. To a marked degree, the New Deal farm program in general, and the rural rehabilitation program in particular, made plains agriculture more economically and environmentally stable. The Resettlement Administration and the Agricultural Adjustment Administration compensated farmers for taking marginal lands out of production. The rural rehabilitation program and the Soil Conservation Service also encouraged farmers to practice soil conservation by field terracing and by raising soil-enhancing crops. After the 1930s wheat growers in the semiarid part of the region lay half their land in fallow to conserve the soil and precious moisture.

On the other hand, in places such as Barnes County, North Dakota, after 1939, farmers may have benefited more in the short term by eschewing the FSA's recommendations for small-scale livestock growing in favor of extensive cash-grain operations. The FSA's long-term agricultural goals meant little to these farm families if they could not take advantage of ample rainfall and better commodity prices. They needed to pay off their debts and garner some profits immediately. Large-scale agriculture and cash-grain operations made much greater profits for North Dakota farmers during the prosperous war years than did the smaller-scale livestock operations for Coffey County farmers. The plains rural rehabilitation program attempted to correct the unsound farming practices among borderline farmers that contributed to the long-term economic, social, and environmental instability of the region. By encouraging its clients to engage in smaller-scale, semisubsistent livestock growing, the FSA tried to make its clients less dependent on outside capital and better adapted to the volatile environment of the region. The capital-intensive, mechanized grain and livestock regimen had led to large debts, huge commodity surpluses, unwise expansion into semiarid lands, and a diminished standard of living among most farmers in the region by 1932.

However, the farm sector still embraced the triad of land, capital, and productivity, despite its historic instability in the plains, as the most profitable and therefore the most stable strategy. Farmers operated their farms as businesses in order to maintain and advance their way of life. Most of them accepted highly capitalized, large-scale agriculture, regardless of its historic instability. This acceptance entailed a gamble of sorts, but one that was in line with the values of the region and adaptable to different conditions. With hindsight, we can see that the logic of this speculative and profitable kind of agriculture presented each county in this study with a different message. A Barnes County client farmer looking ahead one decade in his crystal ball in 1935 would have come to a profound conclusion: Stay on the farm. Wait until wheat prices climb during the war. Convert to full spring-wheat production and rake in the $7,000 in annual gross income that was common by war's end. A Coffey County client would have come to a different conclusion: Stay on the farm. Wait until the economy improves in the early 1940s. Sell off the livestock and farm implements. Move into town or a large city where jobs on construction crews or in war plants paying $45 a week were plentiful.

This part of my study, of the borderline farmers in microcosm, suggests that the plains rural rehabilitation program played a critical role in its client families' lives. It helped sustain them and maintain their farms through difficult times. However, the FSA did not alter the countryside significantly. It kept its clients "farming in place" through grants, loans, and advice. Because of market and environmental conditions, and because of limited funding, the FSA could not transform borderline farmers into larger operators capable of keeping pace with the accelerating trends of large-scale agriculture in the plains. Sustaining these farm families on the land during years of crippling drought and economic depression was by no means a failure. However, during the early 1940s few in the farm sector doubted that increased agricultural production was the farmers' and the country's deliverance. In some areas, such as Barnes County, North Dakota, rehabilitation clients were better placed to take part in the coming prosperity of the war years. However, economic forces during World War II acted to both drive borderline farmers from the countryside and pull them to better-paying jobs in towns and cities across the country. As a result, the FSA was left exposed to the political forces and ideologies that killed it.

1. **The New Beats Out the Old.** During harvest a tractor pulls a binder to cut wheat and bundle it into sheaves. In the 1930s tractors on Great Plains wheat farms could drive farm equipment longer, more efficiently, and more cheaply than ones pulled by horses or mules. (Marshall County, Kansas, July 1927. Kansas State Historical Society.)

2. On the Dole. North Dakota farmers wait uncomfortably for grants from the local Resettlement Administration office. While they demanded emergency aid from the federal government, many plains farmers were uncomfortable with the stigma and paperwork that came with it. (July 1936. Photo by Arthur Rothstein. Courtesy of the Library of Congress, LC-USF 34-5141-E.)

Opposite page

3. Going Over the Books. A farm couple who have taken out a Farm Security Administration loan review their accounts with the FSA supervisor. Clients found that along with the cheap credit, they took on the FSA as an additional partner in their farm enterprise. (Adams County, North Dakota, February 1942. Photo by John Vachon. Courtesy of the Library of Congress, LC-USF 34-64882-D.)

4. Up with the Chickens. Home supervisors actively encouraged farm women to raise food for themselves and to sell in town. Here, an FSA client feeds her flock. (Sheridan County, Kansas, August 1939. Photo by Russell Lee. Courtesy of the Library of Congress, LC-USF 34-64882-D.)

5. Raising Cane. To qualify for a loan, FSA clients had to commit to raising their own feed for their livestock, rather than growing grain for sale. Unloading sorghum cane from wagons for cattle feed was hard and dirty work. (Gage County, Nebraska, October 1939. Photo by John Vachon. Courtesy of the Library of Congress, LC-USF 34-8663-D.)

6. At the Wheel. Usually FSA supervisors encouraged clients to farm with horses or mules to lessen expenses. Occasionally, borrowers used FSA loans to buy tractors. (Sheridan County, Kansas, August 1939. Photo by Russell Lee. Courtesy of the Library of Congress, LC-USF 34-34094-D.)

7. Farmers Take Action. The Farmers Union actively represented the interests of plains borderline farmers and fought cuts to the FSA during World War II. Here, members meet to protest the selling of land to a corporation and to discuss protecting "family size" farms. (Williston, North Dakota, February 1942. Photo by John Vachon. Courtesy of the Library of Congress, LC-USF 34–64786-D.)

8. Nora and Charles Separating Cream, 1945. Farm families separated the cream from whole milk and sold the cream to dairies to supplement their incomes. With World War II many plains farm families began to abandon such small-scale dairying to concentrate on large-scale grain or livestock operations. (Kansas State Historical Society, #WOPO235-36.)

9. Selling Out. By the early and mid-1940s, many borderline farmers in the Great Plains auctioned off their equipment and left the countryside. They were either squeezed out of farming or pulled away by better-paying jobs in war industries. Such auctions were also social events where farm families said goodbye to well-wishers. (Adams County, Nebraska, March 1940. Photo by Al Zimmerman, Nebraska State Historical Society, RG 4289-1.)

6

Politics, War, and the Downfall of the FSA

The Check's in the Mail

Rural rehabilitation in the Great Plains and the United States had a short but controversial life. It was born of the spirit of American liberalism, nurtured by the progressive views of the New Deal, limited by political and financial restraints, and killed by the machinations of interest-group politics. The rural rehabilitation program under the Resettlement Administration and Farm Security Administration grew out of federal efforts dating back to the Hoover administration to buttress sagging commodities markets and distribute feed and seed to drought-stricken farms. However, only after five years of merciless drought, pestilence, and economic travails did the program coalesce in 1935 to address long-term insolvency among borderline farm families.

The New Deal was a reform movement operating within a fundamentally conservative country. Government initiatives like the rural rehabilitation program, as historian Anthony Badger has noted, acted as "holding operations" for distressed classes until the economic recovery arrived. Such programs, Badger said, were circumscribed by the limited "state capacity" of American government. Although the USDA had extensive planning capabilities, the agency lacked the staff, funding, and consent of President Roosevelt, Congress, and the American voters to fundamentally change the countryside. Furthermore, according to Badger, "broker state" politics had emerged by the late 1930s to determine the course and limitations of government actions. The broker state was an extralegal system in which interest groups acted as mediators between government and the American people. Within the farm sector, an "iron triangle" of brokers made up of farm-group leaders, influential members of congressional committees, and USDA commodity experts steered farm policy for the country.[1]

To understand both the support for and opposition to the New Deal in the plains requires a close look at the region's attitudes toward farm programs in general during the 1930s. Relevant to this understanding are not only the views of farmers themselves, but also the opinions of legislators, the press, farm organizations, and other private citizens in the region.

According to Anthony Badger, at its core the New Deal agricultural program attempted the "reform of agriculture through planning in the interests of efficiency."[2] When the New Dealers took office in 1933, problems of gigantic farm-commodity surpluses, massive soil erosion, greatly diminished farm income, and crippling mortgage debt burdened the American countryside. USDA planners wanted to use government bureaucracy to bring farm income, productivity, resources, and credit all in line for the benefit of both farmers and the nation. Plains farmers, however, went along with the New Deal's farm program only so far as it coincided with their desires to keep their homes and property, bring their debts under control, and increase their income.

Plains Opinion and the New Deal

Rural rehabilitation was an important topic in the plains during the New Deal years. However, the farm subsidy program under the Agricultural Adjustment Administration eclipsed rural rehabilitation in terms of both the dollars it pumped into the countryside and the number of farmers affected. In terms of work relief, the Works Progress Administration made more news than the rural rehabilitation program. Clearly, the AAA and WPA transcended the rural rehabilitation program in funding, clientele, and the debate they engendered. In the rural rehabilitation program, however, farmers found a combination of the AAA's attempt to boost farmers' financial security and the WPA's efforts to relieve social ills. Furthermore, the AAA and WPA paved the way for individual farmers and rural plains citizens in general to accept government intervention in their lives. Thus, rural rehabilitation was greatly influenced by the voting patterns and public opinion that formed in response to these New Deal programs.

To truly gauge the reaction to the New Deal, its farm measures, and the rural rehabilitation program, we must survey the plains region at different levels. We begin with a look at voting patterns in Congress and the voting behavior of plains senators during the early and mid-1930s, followed by a survey of opinions from the states' political parties, newspaper editors and the farmers themselves. All these sources reflect the rhetoric on both sides of the New Deal. First, a survey of voting patterns for representatives for Congress between 1932 and 1936 reveals the complexity and volatility of voters in the Great Plains. For example, though North Dakota voters sent Republicans William Lemke and Usher Burdick to Congress during most of the thirties, calling these congressmen "conservative" would be inaccurate. Lemke, for example, was active in the state's Nonpartisan League, which sought to socialize grain elevators, flour mills, and meat-packing companies in the state. During the mid-1930s Lemke criticized the New Deal as too conservative. Burdick, on the other hand, consistently voted as a Progressive Republican and supported many New Deal measures.

During the first four years of his tenure, Roosevelt enjoyed support from plains senators. Within the region the upper house remained solidly in the party of the opposition, the Republicans. During the 1930s only two Democrats served full senate terms in the plains states, George McGill of Kansas and William Bulow of South Dakota. Yet between 1933 and 1936 leading plains Republican senators backed President Roosevelt's initiatives. Long-term Republican senators Lynn Frazier of North Dakota, Peter Norbeck of South Dakota, George Norris of Nebraska, and Arthur Capper of Kansas often supported Roosevelt throughout his first term. Continual drought and economic distress, politics, and a common outlook tied these GOP senators to the early New Deal. Frazier, Norbeck, Norris, and Capper formed their careers in small towns and rural areas of the West and were suspicious of eastern financial interests whom they blamed for the stock market crash. Furthermore, they were troubled by fellow-Republican President Hoover's ineffectiveness in alleviating the anguish of working-class and middle-income Americans. Consequently, these influential plains senators welcomed Roosevelt's forceful leadership in programs that helped the economic and social conditions of farmers, blue-collar workers, homeowners, consumers, and, of course, brought federal dollars to their states.

Roosevelt's consultation with them about New Deal programs, as well as the constant flow of aid coming from Washington to meet the distress of the 1930s, reassured these plains legislators. In 1933 Senator Arthur Capper of Kansas publicly expressed his appreciation for Roosevelt's solicitous aid and called for cooperation between the Republicans and Democrats. The senator also commended the new assistance flowing to his state and Roosevelt's attention to leaders from both parties. He praised the new president for his attention to the farm sector: "I say it is highly creditable both to the vision and the intelligence of the Roosevelt administration," Capper remarked, "that it elected to attempt to restore purchasing power to agriculture first."[3]

After the 1936 election, however, most Republican senators deserted Roosevelt and the New Deal. With the exception of Senator George Norris of Nebraska, they rejoined the ranks of the loyal opposition for various reasons. Roosevelt's court-packing plan of 1937 troubled them. They were also alarmed about the increasing powers the executive branch accumulated in domestic and foreign policy. Western senators in particular felt that Roosevelt, with his new coalition of supporters among southern conservatives, northern labor, and urban voters, ignored them. Finally, the severe 1937 recession alarmed GOP senators and indicated that economic recovery was a long way off.

During Roosevelt's first term, Congress passed a bewildering number of bills to help American farmers. One of these measures, the establishment of the Agricultural Adjustment Administration in May 1933, had a decisive impact on the plains countryside. The goals of the AAA were to cut the mammoth surpluses of crops and livestock and increase farmers' purchasing power. Those who agreed to take part in the program received a payment for cutting back on their production of certain commodities. In the plains, the most important of these commodities were wheat, corn, and hogs. The AAA raised the money for these payments through a tax on food processors.

Subsequent changes in this New Deal farm program angered many plains farmers and their representatives. In January 1936 the U. S. Supreme Court declared the original Agricultural Adjustment Act unconstitutional. In response, Congress rushed through the Soil Conservation and Domestic Allotment Act in which farmers leased land to the federal government which they had used for soil-depleting crops. Two years later, Congress passed the Agricultural Adjustment Act of 1938. The second AAA retained the conservation measures of the stopgap bill and added Secretary Wallace's principle of the "ever-normal granary." Under the second AAA, the government set compulsory production quotas on selected farm com-

modities after two-thirds of the affected farmers agreed to them in referenda. The government lent money to farmers by purchasing their surplus
at somewhat less than the "parity price." Parity was devised to ensure that
farmers retained their purchasing power through the sale of their goods. If
commodity prices rose, farmers repaid their loans and sold their produce
on the market. If the prices remained low, the government bore the loss.
The ever-normal granary was supposed to ensure that farmers received a
fair price when markets were down and consumers had a cheap supply of
food. This farm program remained essentially in place for sixty years.

Although plains farmers initially supported the second AAA, it eventually
became unpopular for two reasons. First, it failed to cut the huge grain
surplus of the time. Second, the USDA used this program to discourage
wheat production by paying off the loans at the legal minimum rate. Plains
wheat growers were therefore left with less acreage and lower income. The
second AAA was so unpopular in Kansas that the bill's cosponsor, Senator
George McGill, a Democrat, was swept out of office during the 1938 election. For this and other reasons, many within the plains farm sector actively opposed President Roosevelt and his farm program in the late 1930s.

From the state political parties and their leadership came a mixed and
changing message regarding New Deal programs. Several Democrats
served as governors in the normally Republican plains states throughout
the 1930s and the early 1940s. These governors, whether Democrat or Republican, backed the relief programs that brought millions of dollars in aid
to their states. Nevertheless, Republican governors criticized this aid. For
example, the Republican governor of Kansas, Alf Landon, supported the
New Deal up until 1935, when he was touted as the GOP's future presidential nominee. Afterward, he criticized Roosevelt for curtailing Americans'
liberties and for his administrative failings. During the 1936 campaign
presidential candidate Landon supported relief while faulting programs
such as the WPA as Democratic political machines.

Despite these criticisms, the political parties of the region saw New Deal
relief programs as necessary. This was especially true for the rural rehabilitation program. Politicians on both sides of the aisle knew that helping
the "deserving" class of farmers get on their feet was popular with their
constituents. Both Republican and Democratic party platforms in the
plains states approved of rural rehabilitation efforts. Some platforms went
beyond assisting insolvent families on their current farms. For example, in

1938 South Dakota Democrats demanded that propertyless farm families be allowed to settle on repossessed land with small down payments and low interest rates.[4] In 1940 both Democrat and GOP state platforms praised rural rehabilitation efforts. That year the Nebraska Democrats lauded FSA loans for allowing Nebraska's "deserving farmer" to continue operating. The platform also called for the "constructive elimination" of farm tenancy through greater landownership. North Dakota Democrats also advocated the FSA's program for helping the landless. Finally, South Dakota Democrats lauded rural rehabilitation, and the state GOP as well commended the program as long as it was "financed by the federal government, administered locally by bi-partisan boards at a minimum expense."[5]

This interparty support was intentional, for President Roosevelt during his first term crafted an image that transcended partisan and class divisions. Roosevelt's rhetoric before the 1936 presidential campaign fostered solidarity, interdependence, and cooperation in the face of disastrous economic conditions. This cast the New Deal as relief and reform demanded by the times. Kansas Republican William Allen White even praised it. In 1934 the newspaper editor expounded on the New Deal: "Much of it is necessary. All of it is humane. And most of it is long past due."[6]

Whatever their political philosophies, people across the region recognized the importance of New Deal benefits. Plains newspaper editors walked a fine line between their personal reservations about the New Deal and the popularity and necessity of its aid. In 1934 a North Dakota newspaper chastised Republican Senator Gerald Nye, a critic of the New Deal. His catcalls were unwelcome, and even dangerous, if North Dakota wished to feed at the government trough. The newspaper stated, "[W]e cannot play poker with the new dealer and expect a pair of aces unless we are at least friendly" to Roosevelt.[7]

One leading newspaper, the *Lincoln Star,* consistently applauded the New Deal and the rural rehabilitation program in the 1930s. The Nebraska newspaper's unfailing message was that this new federal intervention kept the countryside solvent during hard times. The *Star* continually battled critics of the Roosevelt administration during the election year of 1936. The newspaper praised Secretary of Agriculture Henry Wallace who "kept the American farm family intact . . . during years that were as bruising as ever experienced on the farms." Wallace "kept fathers and mothers and children together under the same roof; and he gave them new hope, new faith through sympathetic understanding of the terrific struggle in which they were engaged."[8] Two days later, the *Star* commended the Resettlement Administration's work in Nebraska. Through rural rehabilitation efforts, thousands of farmers who might have left farming were propped up despite criticisms of this government assistance. The newspaper chided, "And who is there who will not say that this is the genuine American way, the good neighbor policy? . . . Is this waste and 'criminal extravagance'? No, it is the Roosevelt Way. It is the real American way."[9]

In October the *Star* cheered those rural rehabilitation clients, aided by bumper crops and soil conservation payments, who were able to repay their loans. "In what condition would these farmers have been had the Hoover program of waiting been continued?" the newspaper asked. Client farmers survived because in Roosevelt they had "a man of sympathetic understanding, who knew that if the nation is to prosper agriculture must prosper, and who had the vision and the courage to take action."[10] However, such support was rare among plains editors, and the *Lincoln Star* was not representative of newspapers in the region. Of nineteen major plains newspapers, only two—the *Star* and South Dakota's *Huron Huronite*—endorsed Roosevelt for the presidency in 1936.

The plains farmers' relationship to Roosevelt's farm policy is of special importance to this study. After President Hoover's perceived inaction from 1930 to 1932, plains farmers were ready for a change. In response to an unscientific mail poll conducted in early 1933 by the Kansas City Chamber of Commerce, 84 percent of Kansas and Nebraska farmers responded negatively when asked whether they wanted Hoover's farm programs continued. Would they favor attempts by the federal government to control agricultural production in order to increase income? Seventy-two percent answered "yes." Finally, an overwhelming 86 percent approved the federal government's refinancing mortgages and other debts at lower interest rates with longer repayment schedules.[11]

President Roosevelt felt compassion toward the plight of the distressed American farm. He also had the political savvy to know farmers wanted a change. Members of both houses of Congress knew this as well. All indications suggest that the original AAA was popular among plains farmers. The vast majority of farmers in most rural neighborhoods voluntarily signed up for the program. They also kicked and screamed in defense of the AAA when it came under fire. Sixty-two Nebraska farmers paid their own way to Washington DC in May 1935 to support the program against critics. Their spokesman, C. B. Steward, defended the processing tax used to fund the AAA payments as "the farmers' tariff." F. L. Robinson, a Buffalo County farmer-stockman, defended the AAA as "the only legislation that we know of that is capable of bringing about parity prices." He continued, "The whole country is interested in getting back our lost prosperity, and all agriculture wants is an equal chance with industry in this fight for prosperity."[12]

Farmers throughout the plains looked upon the commodity checks not as a form of government relief but as their right, just as the industrial East benefited from tariffs. Theodore Alford, reporter for the *Kansas City Star* in 1935, found such sentiments in Kansas. Despite the red tape and government controls on their operations, Alford found many farmers favored making the AAA part of a permanent policy. They looked upon the program as farm "adjustment" rather than control. Therefore, while the government asked farmers to cut back on acreage and production in 1935, they expected to increase production in the future. To the reporter's obvious dismay, the plains farmers had grown comfortable with federal controls and subsidies for their crop and livestock production and had budgeted the AAA payments to supplement their income. Apparently an opponent of the New Deal, Alford searched for chinks in the AAA's armor. He claimed that farmers were "not sold on the soundness of the government farm program, but on the necessity." Still, Alford lamented that "government help in time of distress has been more potent than all the political sophistries on the sacredness of the constitutional rights, the inherent privilege of running one's own affairs as one deems best, and the appeals to the western pioneer spirit of rugged individualism." The reporter closed, "The measuring stick of many farmers has become the same as that of other groups which now are receiving artificial stimulation: 'What do we get out of it?'"[13]

Despite Alford's misgivings, plains farmers accepted "the program," as they called it, simply because it made sense. Farmers eagerly accepted their commodity checks—and the controls that attended them—as one of the strategies for financial security during hard times. In their drive to increase their scale of operations, they took the aid as their due. Hard times called for such changes. However, when the prospects for prosperity returned in the late 1930s and early 1940s, many farmers grew impatient with such controls.

There were signs of returning prosperity after President Roosevelt came into office. Except for dips in 1935 and 1937, plains agricultural income grew steadily, if slowly, during the 1930s. This continual rise kept hopes alive for economic recovery. In addition, farmers knew the environmental disasters of drought and grasshoppers that struck the region were beyond the control of any political party, Democratic or Republican. Finally, the New Deal successfully worked to save commercial agriculture. Particularly when the benefits of the New Deal's farm programs kicked in, farmers felt no need to radically change the system. As current or potential landowners, farmers were fundamentally conservative in their political philosophy. They favored novel initiatives in the 1930s such as moratoria on mortgage foreclosures and government rural rehabilitation programs. However, these were meant to secure farmers' property and opportunity rather than challenge rural entrepreneurial capitalism. Even though they acted cooperatively through the AAA for government checks, farmers wanted to act as individuals using their property to profit from the marketplace.

Another reason for the popularity of the New Deal was that many felt an emotional bond between themselves and President Roosevelt. People of the Great Plains felt discouraged, beaten, and ignored by the time President Roosevelt entered office. Perhaps Roosevelt's greatest feat was to restore Americans' confidence in their country, in their economy, and in their future. Plains borderline farmers desperately needed to boost their morale. Most families were not starving during these years, but they lacked economic security. When they became insolvent, borderline farm families lost not only income but their social status as well. For example, in 1936 one woman replied to a politician's boast that no one was starving in North Dakota. In a letter to the *Fargo Forum,* this woman, signing herself "Mrs. Unemployed," could have been speaking for many borderline farm women: "Speaking for myself I know how it feels to be squeezed out of your farm home, losing all your means for a livelihood together with insurance policies, membership in lodge, church, etc. [and to] settle in town to be promptly classified as a 'no good nobody,' not worthy of consideration."[14]

Throughout the nation in city, town, and countryside, hundreds of thousands of such families—some would call them the "petty bourgeoisie"—had been thwarted in their quest for security and opportunity. Teachers, salesmen, store clerks, and bank tellers who aspired to middle-income status and security now had neither. Photographs from the Great Depression show mature men in three-piece suits, fedoras, and dress shoes standing in soup lines. Franklin D. Roosevelt showed concern for these people in urban and rural areas shunted aside by economic forces.

In response many in the plains had a great interest and affection for Roosevelt. One relief staffer on a trip to North Dakota reported that meetings, sewing circles, and card parties came to a halt whenever Roosevelt spoke on the radio. People gathered in hotel lobbies or wherever there was a working radio to listen to his speeches.[15] Roosevelt appealed to plains residents on the individual and group level. A Miltonvale, Kansas, farm woman described her troubles to the president in a letter. Then she wrote, "Pres. Roosevelt you'll never know how many kisses and hugs you get from the little folks. . . . Well you stay in the band wagon We are with you."[16] Politicians in the region quickly recognized the president's appeal. One attorney wrote James Farley, chair of the Democratic National Committee, in 1936 to report on Roosevelt's campaign stop in Bismarck, North Dakota. When Roosevelt first arrived in Bismarck, crowds were unenthusiastic. But after speaking a few minutes, "he warmed the audience and the result was magic." The attorney closed, "His popularity must be politically capitalized."[17] Both President Roosevelt and the Democratic Party had widespread support in the plains between 1932 and 1936. However, plains voters between the 1936 congressional elections and the 1940 presidential election favored Roosevelt rather than the Democratic Party. In 1940 they rejected Roosevelt, too.

A final reason for the popularity of the New Deal farm programs in the plains was that Republicans presented little competition. If plains farmers found themselves acquiescing to the Democrats' New Deal, one reason was that the Republicans offered no distinct alternative. For example, while preparing for the 1936 presidential election, several Republican leaders offered farm plans of their own. The *Baltimore Sun* noted that each plan offered federal farm subsidies, nebulous soil conservation plans, tariffs to block foreign commodities, and a larger overseas market for American agricultural goods. The newspaper saw in the plans "an absence of anything particularly new" or different from the New Deal program.[18] Their strategy failed to address huge commodity surpluses in the United States and their programs for protection and overseas expansion contradicted each other.

In 1939 *Colliers* magazine observed that the GOP had simply repackaged its farm program to make it more attractive to traditionally Republican farmers. Under the banner of "An American Price for the American Market on the Homestead of the Free," the party offered farmers a two-tier price system for foreign and domestic commodity markets. Most importantly, the Republicans promised to end regulations and red tape with this program.[19] This scheme was similar to the McNary-Haugen Act that President Hoover vetoed, both for its expense and because it failed to address American agriculture's long-term problems. These plans were half-baked and lacked originality. However, the GOP did salvage what farmers liked about the New Deal farm program—commodity supports—and criticized what farmers disliked—red tape and crop controls.

The early to mid-1930s saw a shift in plains citizens' expectations of their relationship with the federal government. The trauma of drought, the collapse in the farm market, and Herbert Hoover's ineffectiveness in dealing with them left the region open to new ideas. During President Roosevelt's first term, respected plains Republicans, state party leaders, the press, and farmers realized the necessity of federal aid and supervision within the countryside. They accepted federal dollars and Roosevelt's leadership because of their timeliness and because the GOP offered no real alternative to the New Deal farm program. However, by the end of the 1930s, fears of Roosevelt's power, mistrust of his political alliance with southern conservatives and urban labor and immigrant groups, the uneven economic recovery, and the unpopularity of the second AAA program turned many in the plains away from the Roosevelt administration. In addition, popular initiatives such as commodity supports and mortgage refinancing had been around long enough and were seen as USDA, rather than New Deal, programs. Plains Republican voters and their representatives realized they could have the new farm programs without the New Deal. However, they also opposed further reforms from Washington, and their support for "emergency" programs such as rural rehabilitation was flagging.

Paternalism, Regimentation, Experimentalism

Understanding the forces that led to the demise of the FSA during the war years demands a review of the ideological opposition to the New Deal farm programs. We will also look at the RA and FSA in particular during the

1930s, both at the grassroots and the state and national political levels. Plains conservatives voiced their opposition to the New Deal's supposed paternalistic and centralizing tendencies, and the Resettlement Administration's "experimental" proclivities. Their fears of and animosity toward an active federal presence in the rural plains helped fuel the demise of the FSA during World War II.

The historian Catherine McNicol Stock addresses these concerns at the community level in her book, *Main Street in Crisis*. She focuses on the reaction of many plains voters to both the trials of the 1930s and the New Deal programs intended to treat them. Stock highlights how the Great Depression, drought, and the New Deal influenced the "moral economy" of "petty producers" in North and South Dakota. A large middle class of farmers and small business owners, she writes, was the backbone of Dakota society when the economic slump hit the plains. Normally, strong values of hard work and thrift combined with community bonds of "neighborliness" to temper the rigors of market capitalism. However, the catastrophic economic depression and drought of the thirties created a "cultural crisis" that revealed the economic, ethnic, and religious divisions of small-town and rural Dakota society. In response to demands from the Great Plains, millions of dollars in relief and agricultural subsidies flowed into the Dakotas. But New Deal initiatives were met with mixed feelings in the region. Although residents were thankful for assistance from the federal government, they were also apprehensive about it. As Stock notes, during the 1930s many Dakotans were ill fed, ill housed, ill clothed—and ill at ease with the New Deal.[20]

One fear within the plains was that the relief policies of the New Deal would drive thousands of "reliefers" into permanent reliance on federal assistance. Conservative publishers, editors, and politicians grumbled that dispensing aid without accompanying responsibility made clients dependent on the government and Washington bureaucrats, and opened them to manipulation by the Democratic Party. This alleged danger came under the term of "paternalism." The *Omaha World-Herald* predicted in 1936 that the New Deal offered a future of "a permanent policy of paternalism that threatens to undermine character and self-reliance and lead great masses of people to depend more and more upon government for their support."[21] Three years later, a South Dakota woman wrote, "I cannot help but feel that this form of assisting people who, for one reason or another, are unable to find work, leaves a good deal to be desired. No matter how long they work on WPA projects, they are, in the end, just as helpless."[22]

Charges that the New Deal was overly indulgent toward the needy were bound to hit the RA and FSA eventually in the plains. The *Nebraska Farmer,* for example, criticized the FSA's active involvement in its clients' farm operations. The magazine's owner, Samuel McKelvie, who sat on Herbert Hoover's Federal Farm Board, complained that the FSA was maintaining a base of hobbled, dependent farmers. In 1940 McKelvie understandably claimed that FSA loan policies stunted clients' growth by curtailing their ability to increase their scale of operations. FSA policies had made Nebraska stockmen run operations only at the subsistence level, McKelvie charged. Instead, client farmers should have been able to build up foundation herds of cattle and hogs that they could readily increase in response to markets, he wrote.[23] Also, that year a Nebraska newspaper criticized the "paternalistic" behavior of the state's FSA staff at a local meeting. According to the *Ravenna News,* the state director's tone toward the clients was tactless and condescending, like that of a stern parent toward errant children. "It was not a sympathetic talk, but cold and relentless. Instead of being helpful, it fostered helplessness. There were men in that room, who 25 years ago, could [have] probably bought all those state men combined," the newspaper protested. During the meeting FSA staffers treated their clients "as foisters upon the government, like beggars on the street, to be carefully and suspiciously watched, in case they tried to make off with something not theirs." The *News* approved the comments of the county's FSA farm supervisor, but claimed that their local rural rehabilitation program "got off to a bad start by leaving a bad taste in everyone's mouth."[24]

The most persistent charge against New Deal farm policy in the plains was that its goal was to take away the autonomy of individual farmers and the entire agricultural sector. One thoughtful plains legislator, Representative Clifford Hope, Republican of Kansas, expressed his misgivings that the crop and livestock reduction within the New Deal farm program stole the coveted freedoms of western farmers. Hope's ideological journey matched those of many plains Republicans. Hope voted against the first AAA, but eventually he approved the voluntary and cooperative aspects of the program and worked diligently to make sure farmers in his district received their commodity checks. After 1936, however, he voiced his misgivings about what he saw as compulsory crop controls. He envisioned a "regimented" agricultural sector at the mercy of domestic consumers as well as political and industrial interests. The results of the New Deal farm program would be, in Hope's words, "more dictatorship and supervision from above."[25]

Some in the plains accepted the increased government involvement in their lives, but only with hesitation. Looking at crop controls, the *Omaha World-Herald* asserted that the countryside would accept them, but "it is a reluctant conclusion which has been forced upon the farmer. They do not relish interference with their industry any more than others," stated the newspaper. "But they are willing to accept the control, regulation—call it what you will—rather than continue to go through what they suffered" in the depression.[26]

Despite the necessity of commodity checks, and despite that controls ultimately came from a majority vote of farmers themselves, the *World-Herald* dreaded the perceived loss of autonomy. Whether the second AAA's production controls came from the government or from majority referenda passed by farmers, they still seemed mandatory. If this was not compulsion, stated the *World-Herald*, what was it? "To a man who wants to market all he grows, a law which in effect forbids him to do so must certainly strike him as compulsory."[27] It was only a short step away from these fears of the regimentation of the entire countryside to the fear of the RA and FSA's supervising their clients' farm operations. Reflecting on the AAA's investigation of crop quotas, the *Topeka Journal* chastised the increased regulation of farmers. The federal government hired new staff "for the sole purpose of visiting the farmers and telling them what to do," and sanctioning them if they refused, complained the newspaper.[28]

New Deal political opponents also made virulent charges against its "regimentation" tendencies. Throughout the plains, the states' Republican Party platforms castigated the New Deal for stealing the farmers' power to run their own operations. In 1934 the South Dakota GOP reproached "the real serfdom into which our farmer population has fallen under the regimentation by Federal authority."[29] That year the Nebraska Republican Party criticized the USDA's management of farmers' herds and opposed "compulsory control of farm production." Future prosperity, according to the Nebraska GOP, came from confidence in the economy. It could not be restored "by the destruction of individual initiative [or] by bureaucratic interference in and control of business, farming, and industry." The Republican platform stated, "We are opposed to any policy that has as its ultimate end the socialization of farm lands, and condemn the policy of the national administration dominating from Washington all business, all labor, and agriculture, including six million farms."[30]

By 1940 influential conservative critics in the plains equated domestic "regimentation" with pernicious trends brewing in Europe. Americans of all political persuasions understood the dangerous totalitarian and belligerent actions of Nazi Germany, Fascist Italy, and the Communist Soviet Union. Though Soviet collective farms and Fascist Blackshirts were not filling up the American Great Plains, both supporters and critics of the New Deal used the global crisis in democracy of the 1930s to smear their opponents. The Republican Party platforms in the plains in particular equated centralization at home with dictatorship overseas. For example, in 1940 the South Dakota GOP clothed itself in heated rhetoric to attack the New Deal when it proposed "a constitutional and constructive approach to the nation's problems by reopening the gates of opportunity." The New Deal would only lead to "social revolution and ultimate enthronement of either a fascist or a communist dictatorship."[31] The same year the Nebraska GOP reaffirmed its allegiance to the principles of the Founders of the Republic. The party then stated its opposition to "the evils of the new deal" and demanded the "repeal of emergency dictatorial powers and the restoration of the lawmaking function to congress."[32]

Critics of the New Deal in the plains aimed their most pointed barbs directly at the Resettlement Administration and specifically at its head, Rexford Tugwell. They called him "Rex the Red" and chided his rural projects as chimerical. Tugwell exacerbated the conflict through controversial statements made while head of the RA. For example, in October 1935 he upset Democratic regulars by promoting a new "Progressive Party," resembling a Farmer-Labor party at the national level. Tugwell's left-of-center ideology and association with novel government initiatives made him a target for plains Republicans. Because GOP members knew President Roosevelt and his New Deal were popular throughout the country they made Tugwell their lightning rod for New Deal activism in the rural plains. During the 1936 presidential election, Samuel McKelvie in the *Nebraska Farmer* berated Tugwell for his "Russian theories of government." In the same issue of the magazine, a Republican Party advertisement asked farmers chill-ingly, "Do You Want to Be a 'Hired Man' for Wallace and Tugwell?"[33]

From the Dakotas came amused and heated opposition to Tugwell. In North Dakota, the editor of the Dickey County Leader pondered, "I can't be-lieve that the President is a radical. He may let Mr. Tugwell, et al, play around at left winging, but if he sees it isn't suiting the folks so well, he turns the hose on the lads."[34] In 1936 the North Dakota Republican Party chastised Tugwell from the party platform, calling for his removal for his "communistic farming ideas."[35] Even though Tugwell left the RA in 1936, criticism of him continued. In October 1941 the Sioux Falls Argus Leader chided an experimental farm settlement set up by the RA outside the city. The South Dakota newspaper recalled that Tugwell and others like him "used the power and funds entrusted to them to further their socialistic theories." Tugwell, according to the Argus Leader, was "a parasite on any-body's payroll."[36]

Decades later, Tugwell himself reflected on the idea of resettlement and its unpopularity. He recalled that the RA's objective was to address rural underemployment and massive soil erosion. Tugwell confirmed that crit-ics lambasted the plan's alleged regimentation, that soil conservation values had not been adequately introduced to the American public, and that RA clients were seen as undeserving. Perhaps most important, as RA head he neglected to cultivate congressional goodwill. While he got along with Senator George Norris of Nebraska, Tugwell alienated Senate heavy-weights such as Harry F. Byrd, Democrat of Virginia, whose animosity to the RA and FSA proved fatal to rural rehabilitation during World War II. Tugwell admitted that his style compromised the effectiveness of the Re-settlement Administration: "I made the mistake of attempting too much at once—of undertaking battles on more fronts than were necessary or could be successfully fought; and once engaged, I could not or would not with-draw. So I would become not so much a symbol of progressivism as a kind of crackpot radical."[37]

Rexford Tugwell drew such criticism because he showed little regard for established American economic goals of unrestricted business competi-tion, unencumbered property rights, and individual entrepreneurialism. But rather than favoring socialized farm ownership, as his critics charged, he wanted a collective and cooperative system to manage the nation's vast resources. In his defense, during the 1930s Tugwell and other agricultural planners saw the extraordinary difficulties of the rural plains and offered extraordinary solutions. They attempted to use government planning to counter the economic, social, and environmental problems in the region. Projects for farm resettlement and land retirement were only experiments in a region demanding answers. If these programs seemed "visionary," at least they were attempts to solve serious problems. Furthermore, plains farmers themselves showed little regard for the supposed sanctity of free markets and "rugged individualism," instead favoring government help to ease the burdens of landownership during the 1930s. They embraced co-operative crop controls and foreclosure moratoria to save their farms. Given farmers' demands for radical solutions, Tugwell and the Resettlement Administration were not as revolutionary as their critics contended.

Still, it was only natural for New Deal opponents to attack Tugwell and the Resettlement Administration. In the face of the Roosevelt juggernaut of popularity, adversaries used available ammunition against the New Deal. Tugwell seemed to his detractors to question the principles of individual entrepreneurialism and property rights that animated the settlement of the plains. In addition, the resettlement and land retirement projects were easy prey because they were expensive and conspicuously unique in the region. The idea of government acting to *remove* farm families and *withdraw* land from production must have seemed positively foreign to the region, since the state and federal governments had been working seventy years to do the opposite.

As the 1930s ended in the plains, there was waning support for rural rehabilitation programs. The economic depression that shrouded the country for a decade seemed unabated, and many farmers in the region had had their expectations for prosperity foiled since the 1920s. Critics reviled the second AAA for limiting farmers' freedom to expand production and their income. The Resettlement Administration had reaped a legacy of controversy which it passed on to its successor, the Farm Security Administration. Plains voters increasingly put Republicans in the House. Finally, the press, Republican Party, and local agricultural officials in the plains were hostile to expanding the New Deal's involvement in everyday life. The New Deal was acting essentially as a "holding operation" for the social and economic status quo until prosperity returned. The plains farm sector came to accept government intervention in such areas as commodity supports and mortgage relief, but support for "emergency programs" dispensing work relief and rural rehabilitation lasted only in the face of devastating drought and economic depression. When the markets and rains returned, as they did in the 1940s, conservatives and moderates alike questioned the necessity of the rural rehabilitation program.

The Plains Farmer at War

Six weeks after the *Argus Leader* berated Rexford Tugwell, the United States was at war with Germany, Italy, and Japan. At the time the plains and the farm sector were not prepared for their role in the great global conflict. Ideologically, the region wanted foreign markets but not intervention in foreign wars. Plains senators and representatives followed the passionate isolationist sentiments of their constituents. Like most midwestern lawmakers, they voted consistently against efforts to build up American military power, like the move to establish a peacetime draft in 1940.

The nation's primary agricultural problems still remained from the Great Depression: underemployment, substandard living conditions in the countryside, soil erosion, flat commodity prices, and huge produce surpluses. Consequently, the state and federal farm programs continued as they had during the New Deal. Most farm program administrators still operated under a system meant to curb production, cut soil erosion, and keep underemployed farmers in the countryside rather than moving into town to live off county relief. Therefore, except for the mammoth grain surfeit, the American farm sector was unprepared for the demands of global warfare. Between spring 1940 and Pearl Harbor, while the nation's industry converted to wartime production, America's agricultural policies remained much the same.

But transforming a depression-battered countryside to a war economy was extraordinarily difficult. First, it was hard to coordinate and mobilize millions of American farmers. Many were reluctant to expand production unless the government underwrote the risks. Second, over the last two decades, farmers had learned that more production led to surpluses and lower prices. New Deal policies meant restricted output, and many had learned to live with this. Finally, the federal government struggled to address not only the needs of American farmers but also consumers, both domestic and foreign.[38]

Even during the war the New Deal farm program remained intact. In May 1941 Congress guaranteed farmers raising wheat, corn, and other key commodities at least 85 percent of parity. In January 1942 Congress passed the Emergency Price Control Act to restrain inflation in food prices. The act set a price ceiling of farm commodities at 110 percent of parity. Congress also encouraged the American farmer to raise more livestock, vegetables, and oil-bearing crops such as peanuts and soybeans. Blessed by favorable weather, plains farmers rose to the challenge by incredible increases in grain and livestock production. Despite having 33,000 fewer farms in 1945 than at the beginning of the decade, plains farmers enlarged their tilled cropland by 17 million acres—a 21 percent expansion. In five years, plains farmers increased their annual harvest of wheat by 173 million bushels and their harvest of corn by 356 million bushels. Plains stock raisers doubled their production of hogs and expanded cattle herds by nearly two-thirds between 1940 and 1945.[39]

Plains farmers filled the nation's larders but were often discontented. Despite the popular notion of an America united in war, the actual environment in the plains was one ill at ease with the old New Deal and the new demands of warfare. Plains farmers and their spokespersons vented their frustrations with wartime price controls and supports, inflation, and labor and machine shortages. Even before Pearl Harbor some American farm representatives were anxious, fearing that agriculture would be treated as a second-class industry in the nation's preparations for war. In 1940 L. J. Taber, the national head of the Patrons of Husbandry (the Grange), warned a Broken Bow, Nebraska, audience of the dangers of the war economy. Agriculture hadn't received "the consideration that it deserves," Taber protested. Labor and businesses received "first consideration while agricultural problems are being forgotten."[40]

Farmers repeatedly complained that after years in a lagging economy, they still lacked the unfettered markets they desired. The American economy during World War II was not a free-market economy set by supply and demand. Rather, the federal government controlled such key factors in agriculture, manufacturing, and transportation as prices, supplies, wages, and working conditions. For the nation at large, the fruits of the "war prosperity" were elusive. In fact, many Americans were not truly out of the Great Depression until the postwar economic recovery. The war years were a traumatic time when the American economy shifted from a depressed, partially managed economy under the New Deal to a booming, fully managed wartime economy. New Deal regulations had disturbed many farmers during the economic slump, but now they felt thwarted by the red tape and administrative foul-ups that came with rationing and price supports during the potentially prosperous war years.

Everyone from the plains governors to its farmers was anxious that wartime shortages cut into farm production. A midwestern governors' conference in 1943 warned of a crisis in agriculture: "[U]nless we have more manpower and more machinery the food production cannot be maintained. Already serious losses in crop and animal products are imminent; there can be no delay."[41] Elizabeth Chitty of Bigelow, Kansas, wrote President Roosevelt in 1942 that "this exploitation of farming has been causing hundreds of people to abandon the industry." Extreme farm labor shortages, Chitty claimed, meant that 150 acres of her land lay idle. She and her husband had two farm workers, but the couple lacked "the least assurance that these two men will stay on this job when they can earn many times as much on a defense job and work only eight hours a day." Farmhands went AWOL. Around the region they used the free market for their own good and deserted the countryside for opportunities in towns and cities. The war economy wasn't meeting farmers' expectations, and they were fuming. Chitty grumbled that while farmers faced shortages of labor, mechanized cornpickers, and sugar for canning, urban industries seemed to have all the resources they needed. She attached five sale notices of farmers who were leaving farming to her seven-page letter.[42]

Labor shortages and the commodity supports program stymied Dakota farmers. One South Dakota farmer complained, "I am doing the work of three men for one day's pay, while the laborers they've taken from us [for defense plants] are getting three days' pay for a third-of-a-day's work."[43] In 1943 the Democratic county chairman for Lyman County, South Dakota, noted that inconsistent government pricing policies in 1943 deeply upset local hog raisers. The previous winter the USDA encouraged them to expand their hog production. In the meantime, pork prices fell. Since there was an effective ceiling but an insufficient floor on pork prices, many farmers lost money when they sold their herds. "Livestock men," the chairman wrote, "are very much discouraged with what they call a raw deal from the government."[44] Plains farmers were still better off than they were during the depression. However, pressures of war and the loss of potential income left them restless and receptive to GOP attacks on the Roosevelt administration. As one North Dakota attorney noted in 1943, "[T]he farmers of the west have never made as much money as they have now . . . but they are still made to feel dissatisfied by propaganda and [by] such restrictions as are put upon them, and their products [from Washington]."[45]

The American countryside was rapidly and haphazardly shifting from a depression economy into full wartime production. What to do about two precious commodities—farm labor and farmland—was a recurring debate. This was an era of labor-starved larger farms with higher productivity. Yet there were also many smaller, less-productive farms that "underemployed" both the labor of its farmers and the resources of the land. So, while some plains farms were short of labor during World War II, the USDA estimated that 600,000 American farms were so small or infertile that their operators would better serve the country if they went to work in war industries.[46]

The FSA and the Farmers Union at War

It was in this atmosphere of higher agricultural profits, labor shortages, and demands for fewer government controls that the FSA came under attack. By 1943 the FSA, along with other New Deal programs, was a casualty of wartime necessities and changing political, economic, and agricultural trends. Like the FSA, the Works Progress Administration was begun in 1935 in response to the vast underemployment that was crippling America. Despite efforts to show their necessity during wartime, both these agencies were killed by cuts in their funding between 1942 and 1943.

Political conservatism did not capture but simply reclaimed the Great Plains by World War II. The social activism of the New Deal declined in the late 1930s as President Roosevelt turned to problems overseas. In the face of crises abroad, Roosevelt began nominating Republicans for cabinet posts to prepare for a bipartisan war effort. Plains voters also turned increasingly against the Democratic Party and the president himself. After the 1936 election the number of Democratic representatives from the region fell to predepression levels—only three out of the region's sixteen congressional seats remained Democratic. During the war plains voters elected *no* Democrats to the House. Plains support dropped for both the New Deal and President Roosevelt in elections. The region supported Roosevelt during the presidential elections of 1932 and 1936 as much as or more than the country as a whole. However, during the 1940 and 1944 races plains voters only gave him about 42 percent of the vote, 10 percent less than the overall American vote. With the economic crisis subsiding at home, plains voters returned to the GOP. They also seemed to give less support to "welfare state" programs like the WPA and FSA.

But the rural rehabilitation didn't go down without a fight. In the plains its main proponent was the Farmers Union. Its primary opponent at the national level was the Farm Bureau. Like the political parties, both sides launched smear campaigns using wartime rhetoric against each another in their struggle to determine the future of rural America. Supporters of rural rehabilitation and the Farm Security Administration at the national and plains level were varied. Between 1942 and 1943 at the national level, labor organizations, churches, and the Farmers Union office opposed further cuts in the agency's budget. Within the plains, local businessmen, individual farmers, churches, labor unions, and most state Farmers Unions supported the FSA. The composition and the power of these groups relate both the strengths and weaknesses of the FSA's appeal.

At the national level, the FSA attracted urban-based New Dealers, such as religious and labor groups, for varying reasons. The most consistent supporter of rural rehabilitation among religious groups was the Catholic Church. A leading voice of the Church's concern was the National Catholic Rural Life Conference, based in Des Moines, Iowa, and first convened in 1923. Despite their urban image, one-fourth of all Catholics in the United States lived in rural areas during the 1930s. The conference leadership, including men such as Reverend Edgar Schmiedeler of Kansas and Bishop Aloisius Muench of Fargo, North Dakota, was concerned that the Great Depression was driving these Catholic farmers into becoming a rural proletariat. While interested in its efforts to restore the buying power of rural America, many Catholic agrarians were suspicious of the New Deal farm program. Members of the conference feared New Deal programs were "regimenting" farmers and driving them off the land. They also believed, with good cause, that the federal government was more interested in economic recovery than in rural economic and social reform. Despite these reservations, and because the FSA addressed concerns that closely matched their own, the Catholic Church was a leading proponent of the rural rehabilitation program.

When the FSA came under attack during World War II, members of the Rural Life Conference defended the agency. Monsignor Luigi Ligutti, its executive secretary, praised the FSA in 1943 for having done "a splendid piece of work among farmers in general and especially among the poorer farmers. . . . The scuttling of the f.s.a. or the hampering of its work is, in our opinion, equivalent to giving aid and comfort to the enemy."[47] At the regional level, the Catholic bishops of Bismarck, North Dakota, and Concordia, Kansas, wrote legislators in Congress on behalf of the fsa.[48]

The Congress of Industrial Organizations (cio) also threw its support behind the fsa. After forming in 1934, the cio had worked consistently to organize mass-production industrial workers into a union. Led by the fiery head of the United Mine Workers, John L. Lewis, the cio had close ties with the New Deal, which fostered laws protecting the rights of organized labor. At the beginning of World War II, the union was a powerful voice with a membership of 2.6 million workers. However, anti-union pressure from conservative business and political interests was endangering the cio's power on the shop floor.

Faced with efforts by business and conservative opponents to curb New Deal labor legislation, the cio cast about for potential allies such as the Farmers Union and borderline farmers aided by the fsa. Although organized labor and farmers had not been especially friendly to each other, the cio lent its voice in support of the fsa program. In 1943 the labor union charged that a "Hunger Bloc," supported by "big agriculture" and the Farm Bureau, planned to "assassinate" the agency. The cio contended that big agriculture was determined "to deny assistance to small farmers and, taking advantage of the war needs of the nation and its allies, to extend commercialized agriculture and the plantation type of farming, both dependent upon a plentiful supply of cheap and docile labor recruited from among the dispossessed farm families."[49] Within the plains region, union locals contacted their representatives, urging them to stop further cuts in the fsa budget.[50]

Throughout the region businesses backed the rural rehabilitation program, which had profited them. In Nebraska, businessmen who depended on fsa clients' business praised the loans in 1943 for transforming insolvent farmers into stable ones. An abstractor in Kearney, Nebraska, noted that there were some clients "who have not been sincere in their asking for this rehabilitation, and in such cases have made failures of themselves." However, most clients were paying off their debts.[51] Businessman-landlord Rex Henry of Fremont, Nebraska, also commended the fsa for rehabilitating his tenant. In 1940 Mr. Henry was set to find a new tenant to replace the debt-ridden man renting his land. Henry recalled the tenant "was discouraged and was not keeping the place up." After receiving an FSA loan, the renter had paid nearly all his debts. Henry complimented the FSA for doing "an outstanding job" in the Fremont area.[52]

Furthermore, despite the improving economy, thousands of families still required aid. For example, Mrs. Wilks Harper of Larned, Kansas, had been on and off relief since 1933. She wrote in 1942 that she could not provide for her six children and that the army declined to take her husband because of his poor health. The family still wanted to help the American war effort through farming, she claimed.[53] North Dakota farmers during the war expressed their gratitude for the rural rehabilitation program. Martin Jacobson of Renville County wrote that in 1935 he was heavily in debt and stood close to losing his farm. With $2,700 in rural rehabilitation loans, planning, and the "utmost courtesy," the FSA helped him pay off his debts. Hilda and Joseph Metcalf of Fort Rice told a similar story. Although they were not large farmers, with the FSA's assistance they grossed $4,000 during the 1942 crop year. "If we are to keep this a free country," the Metcalfs closed, "give the little fellow a chance, as we are all entitled to a way of earning our living, and also the freedom which we enjoy by being self-supporting."[54]

Some in North Dakota waxed militant in defense of the FSA. The *Bottineau (North Dakota) Courant* charged that the National Association of Manufacturers and a "farm bloc" led by the Farm Bureau sought to kill rural rehabilitation and thus provide cheap labor for industry and larger farmers. Then the *Courant* went into high gear. While the Farm Bureau "accuses the Government of using the war as a cloak for carrying out social reform, we find that the farm bloc is trying to consign 3,000,000 farmers to the chain gang. Using the war as a cloak and a page from *Mein Kampf* as their guide, these would-be dictators are themselves plotting to take control of rural America and transform free farmers into wage slaves under the lash."[55] Wartime labor conditions and shortages motivated the FSA's opponents. There is no data available to prove that large industrial and agricultural interests were set on converting rural rehabilitation clients into ill-paid powerless members of the urban and rural workforces. However, the Farm Bureau and its allies were clearly concerned about government and labor union attempts to organize farm workers and smaller farmers, for fear of rising labor costs.

Others in the plains saw the fight over the FSA as a struggle between wealthier and smaller farmers. Attorney Robert Brower, also a vice president for a Fullerton, Nebraska, bank, attested that rehabilitation clients were repaying their loans. In a letter to Nebraska congressman Karl Stefan, Brower argued that debate over the FSA's future was part of "the larger political fight, between the representatives of larger farming interests and the smaller farmer," especially so in the American South. The attorney claimed that rural rehabilitation clients were essential to increasing agricultural production for wartime needs. "It is from these poorer farmers who have not the means to operate alone that any increase in production at this time can be expected, the others are operating full tilt now anyway." Representative Stefan, a Republican, replied with a marginal tribute. "In the early days," he wrote, "FSA embarked upon a lot of collectivist activities." Yet in Nebraska, the agency "worked along the lines of which you and I approve."[56]

President Roosevelt joined in the FSA's defense. In summer 1942 he reaffirmed the principles of fair commodity prices and organizing farm production for the war effort. Although recent FSA allocations from Capitol Hill were inadequate, wrote Roosevelt, they were better than the cuts urged by "certain selfish and power hungry groups." While billions had been allotted for the conversion of industry to war, the president noted, little had been done to transform American farmers to a war footing.[57]

Roosevelt was responding to a letter sent to him by the Farmers Union, the Ohio Farm Bureau, and labor and religious groups. The groups praised the FSA for expanding clients' production during the war and claimed that attacks came from interest groups committed to the "high-price-through-scarcity concept." With further cuts in the FSA, the letter continued, "Hitler and his Axis partners must be gloating at the headway thus far made in the attempt to slash these funds." The future of FSA meant a great deal to the future of agriculture in the United States, the signatories asserted. The fate of the FSA was "a question of deep concern to labor and to all elements in our population dedicated to the democratic way of living."[58]

As noted, at the national level the most vocal nonfarm supporters of rural rehabilitation were urban labor and religious groups who allied with FSA staffers to keep New Deal activism in the countryside alive. At the national level and within the Great Plains, the Farmers Union was the leading farm group supporting rural rehabilitation, and it lobbied on the FSA's behalf. In 1940 Kansas-born James Patton took over the presidency of the organization, where he labored to bind the state branches more closely to the national office. Before the World War II, 31,000 plains families belonged to the Farmers Union, comprising a third of the nation's family memberships. Nebraska and North Dakota led the region in memberships, with 11,300 and 10,500 member families, respectively. Nationally, 84,000 families belonged to the Farmers Union in 1940.[59] While the Farmers Union was a powerful farm group in the plains, its influence at the national level paled in comparison to that of the Farm Bureau.

There is no data on farm organization memberships among rural rehabilitation clients or farmers by income level. However, the rhetoric on behalf of small, individual rural entrepreneurs probably appealed to borderline farmers and FSA clients. Faced with the juggernaut of the increasing scale of plains agriculture, these farm operators saw in the Farmers Union a kindred spirit. The Union used the rhetoric of an economic underdog, the tool of the farm cooperative, and the ideal of an economic democracy to place smaller farmers on an equal plane with larger farmers and powerful agricultural interests. No doubt many rural rehabilitation clients and borderline farmers were Farmers Union members.

The Dakota and Kansas Farmers Unions had tied themselves closely to the New Deal by 1940. Emil Loriks, the head of the Farmers Union in South Dakota, served as the state director of the FSA from 1940 to 1941. In 1942 Loriks became regional director for the FSA in the plains, replacing Cal Ward of Kansas, who had also been the head of his state's Farmers Union. The North Dakota Farmers Union formed an especially close relationship with the rural rehabilitation program. The Union even controlled FSA appointments in North Dakota. Farmers Union representatives in North Dakota worked closely with state FSA staff to promote grain cooperatives. For their part, FSA supervisors encouraged rural rehabilitation clients to borrow the $25 required for stock in the Farmers Union's grain terminal association.[60]

The FSA needed such alliances. Beginning in 1942, Congress increased its scrutiny of the FSA and slashed budgets for the agency. The exact level of support for the FSA among plains representatives and senators cannot be found. However, after 1942 only two plains legislators, Senator William Langer and Representative Usher Burdick, both North Dakota Republicans, took the floor of the Capitol to praise the FSA.

Therefore, squeezed between astronomically high war budgets and the perception that rural rehabilitation was no longer needed, the FSA depended on allies in the Farmers Union to fight further budget cuts. As Congress deliberated on federal spending, James Patton and the National Farmers Union demanded increased spending for rural rehabilitation. In February 1942 Patton railed against "certain reactionary groups" who "seized the opportunity to launch an all-out offensive" against the FSA under the guise that it was unneeded during a war. Patton praised the FSA for undertaking issues such as rural poverty and high tenancy rates. He also lauded the FSA for its role in feeding America and the Allies while fostering democracy. Patton said before a national radio audience, "What terrible irony it would be if for some incomprehensible reason we could achieve military victory only at the sacrifice of our small farmers, the greatest source of our spirit or independence and freedom." Patton believed cuts for the FSA budget were led by "reactionary interests to use the war emergency as a camouflage—a camouflage, behind which these interests seek to systematically destroy the gains made in social and agricultural legislation over the past ten years."[61]

One month later Patton commended FSA clients for increasing food production for a world at war. He wrote that the nation's "dirt farmers" knew that opposition for aid to "family-type farmers" came from powerful agricultural concerns who received vast government subsidies. "They desire to control agriculture's war policies in their own selfish interest," he wrote, "just as some elements of big business have placed self-interest ahead of victory and public need."[62]

Senator Harry F. Byrd of Virginia, a conservative Democrat, spearheaded cuts in FSA appropriations. Senator Byrd mounted intense opposition to his party's labor, spending, and social legislation from President Roosevelt's time to Lyndon Johnson's. A champion of fiscal conservatism, states' rights, and congressional prerogative, Senator Byrd was a staunch opponent of the New Deal. As head of the Senate Finance Committee, he offered consistent opposition to the Roosevelt administration. Beginning in 1938 Byrd worked to constrict the FSA's budget. He was stymied until the war years by Democrats in the House and Senate from across the country, save for the Northeast.

Eventually Byrd and his allies won converts against the rural rehabilitation program. The Virginia senator used his chairmanship of the Joint

Committee on Reduction of Nonessential Federal Expenditures, which investigated government programs remaining from the New Deal years, to attack the FSA. In 1943 Senator Byrd called the FSA "wasteful" and the "most disregardful of the true interest of the people and of . . . low income farmers."[63] The Farmers Union national office struck back at Byrd and his committee for slicing the FSA budget during a "witchcraft trial." It claimed Byrd was exacting retribution against the FSA for not carrying out the interests of plantation owners. The planters, claimed the Farmers Union, treated FSA clients as "human workstock" with a "dismally small share in American Democracy."[64] Despite the inflammatory rhetoric, the Union had a point. Senator Byrd represented southern planters' interests. As the owner of a large-scale apple orchard business, he saw government regulations increasing his operating expenses. Furthermore, since one-fifth of the FSA's clients were African Americans, the southern Democrat may have fought the FSA for keeping small black southern farmers out of the farm labor pool.

Plains organs such as the *Kansas Union Farmer* reflected the national Farmers Union in their defense of the FSA in 1943 and 1944. The publication praised rural rehabilitation loans for increasing smaller farms' productivity during wartime.[65] Like the national branch, the Kansas Farmers Union labeled the FSA's enemies as large southern plantation owners, industrial interests, and the Farm Bureau. The Kansas Farmers Union criticized the Bureau as one of the large business interests oppressively crowding out borderline farmers. "Obviously," the *Kansas Union Farmer* lamented, "the Farm Bureau won't be satisfied until all the land is owned by big commercial interests and tenants are forced to work for them with no hope of becoming owners of the land they till."[66] Woe to anyone who criticized the FSA. In early 1943 Congressman William Lambertson, Republican of Kansas, castigated the rural rehabilitation program, saying that clients' labor could be used better elsewhere. The statement was economically tenable, but politically ill advised. The *Kansas Union Farmer* shot back that Lambertson was in "complete accord with large, commercial type agricultural interests." Furthermore, the congressman had shown his "complete inadequacy as a representative of the farmers of his district."[67] Despite the furor his statement created, his district reelected him the following term.

Overall, plains Farmers Unions supported the national FSA and fought against further appropriation cuts for the rural rehabilitation program during World War II. An important exception, however, was the Nebraska Farmers Union. Its official publications, edited by state Farmers Union president Chris Milius, had little sympathy for the federal agency and championed its version of free enterprise, higher grain yields, tax reform, and purchasing, sales, and credit cooperatives as the true solutions for the problems of the countryside.

The Nebraska Farmers Union recognized the problems that troubled the rural plains in the 1940s. However, it also opposed government involvement to stem these problems. The official newsletter reflected fears of "stateism" and of dictatorship through concentrations of power of all varieties. Above all, to this conservative state farm group, the free market was the solution the farmer needed. The Nebraska Farmers Union attacked the rhetoric with which its associates smeared the FSA's opponents. Not all the FSA's adversaries were "reactionaries or in league with exploiting landlords or land monopolies." Rather, Milius wrote, many were apprehensive about the growth of government power through the FSA. The rural rehabilitation program threatened the countryside's liberty. According to the *Nebraska Union Farmer*, "The ostensible purpose of the FSA is laudable, but its methods are subversive of freedom and upstanding independence." Rural rehabilitation was "a statistic [*sic*] program that heads toward bureaucratic dictatorship. We do not want to free farmers from exploiting landlords and commercial interests only to make them subservient puppets of the politicians."[68]

Above all, the Nebraska Farmers Union feared controls or limitations on farmers. As the elected leader of the state farm group, Chris Milius expressed his hopes, despite the experience of the past two decades, that unrestrained agricultural productivity, aided by the Union's cooperatives, ensured both profits and freedom for farmers to run their operations as they pleased. The interventions of the FSA's rural rehabilitation and tenant purchase program, in Milius's eyes, threatened farmers' mutually supporting opportunities and freedom.

During the 1930s concern for the smaller-scale farmers, desire for an equitable rural economy, and fear of large agricultural business interests fit the times. After America entered the war, however, farmers were more concerned with issues such as government price supports, inflation, and acquiring the labor and machinery to run their operations and garner profits. Many farmers in the Great Plains still lived on the margins of poverty. However, many more took advantage of high commodity prices and abundant rains to pay off debts and purchase comparatively inexpensive land. Farmers who survived the Great Depression now looked beyond survival to expanding their scale of production.

Conditions for farmers were changing. Yet, as the wartime economy accelerated agricultural costs and productive demands, the FSA could have played an important planning role. For those borderline farm families who remained on the land, the rural rehabilitation program could have worked to maximize their yields and help them move into middle-income status. For those migrating into towns and cities to take war industry jobs, the agency could have assisted in the transition. Unfortunately for the FSA and its clients, because the specter of massive rural poverty diminished with the wartime prosperity, the agency's existence depended less on the concern for the welfare of the nation's borderline farmer and more on influential national power brokers, the Farm Bureau in particular.

The Farm Bureau and the Defeat of the FSA

During World War II the rural rehabilitation program seemed doubly irksome to its opponents. First, during a period of high demand for farm commodities and abundant rainfall, the program seemed superfluous. Second, farmland and farm labor were at a premium. To spend taxpayers' money on a program to keep farmers on small farms seemed an outmoded and unbusinesslike act of charity, and it also appeared to suspend the traditional rules of market competition.

At the national level, the country's leading urban and rural business organizations opposed further funding of the FSA. Within the plains region, certain disgruntled clients, members of the farm press, and more substantial farmers supported cuts for the agency. Overall, powerful interest groups worked to squeeze rural rehabilitation out of existence. Criticism against certain New Deal agencies had a broad base among American conservatives. Increasingly during World War II, demands for individual profit eclipsed the reforms embodied in the New Deal. The wartime business ethos equated profits with the war effort. In this environment, business interests disparaged New Deal goals and class rhetoric. Advertising from the era, for example, celebrated an "American way of life" marked by free enterprise and more consumer goods without organized labor strife. During the 1940s, therefore, larger agricultural and urban business interests simultaneously waged a war against the Axis powers and New Deal constraints on their operations.

In the probusiness war years, groups seemingly as disparate as the U.S. Chamber of Commerce and the Farm Bureau joined in common cause against the remnants of New Deal activism in the countryside. Although the Chamber of Commerce was primarily concerned with urban businesses, the organization opposed the Farm Security Administration for three reasons. First, as a representative of large American business, the Chamber favored cutting government spending on behalf of farmers on the edge of poverty. Second, the business group probably believed that the rural rehabilitation program worked to organize and keep hundreds of thousands of farmers on their farms, which constricted the industrial labor pool and boosted wages. Finally, members of the Chamber feared that the FSA would become the "union" for smaller farmers. Therefore, in 1942 the Chamber of Commerce criticized government efforts to transform America's borderline farmers and chided what it saw as the FSA's too-expansive mission, its permanence, and the centralized federal efforts of county welfare activities. Fundamentally, the Chamber castigated the FSA's attempt to mediate market forces during wartime. The agency's goal was to assist farm families on the margins between solvency and failure, but the Chamber questioned continuing a program "which may interfere with occupational adjustments within agriculture and between agriculture and other industries." Because of wartime prosperity and labor shortages, the Chamber discouraged "subsidizing inefficient farmers."[69]

Conservative farm press representatives and larger farmers in the plains also opposed the rural rehabilitation program during the war. For example, in 1943 the editor of the *Kansas Farmer*, Raymond Gilkeson, wrote his employer, Senator Arthur Capper, that the FSA was "an expensive luxury that is doing everything in its power to perpetuate itself." Gilkeson charged that FSA home and farm supervisors were simply duplicating the work of county farm agents and home demonstration agents of the extension service. The editor doubted that farmers would need the FSA after the war, so keeping it alive for the duration was an unnecessary burden. Gilkeson thought that the end was near for the agency. "While some folks will criti-cise [sic] elimination of FSA now," he wrote, "I feel sure the majority would favor it."[70]

Certain farmers also opposed funding for the FSA. Neal Haskell of Lau-rel, Nebraska, testified to a congressional committee in 1943 on the pro-gram. Haskell ran a large 480-acre operation and was also the vice pres-ident of the state's Federation of County Taxpayers League. The farmer called local cooperative credit programs under the FSA "entirely a political football." Regarding the FSA clients who still needed assistance, despite wartime prosperity, Haskell contended "if they cannot operate on their own ability, on their own profits, it is not right they should be continued to be subsidized by the Federal Government any more." Haskell continued, "If these farmers that they are helping cannot paddle their own canoe now, Lord help them when things get tough again."[71]

Both Gilkeson and Haskell contended that the FSA was expensive and of questionable value during the wartime economic boom. In truth, in 1943 the national rural rehabilitation and tenant purchase programs cost a total of $49 million. Although this was a large amount at the time, it was only 5 percent of the total spent on USDA programs. On the other hand, few at-tacked the federal land banks, which were federal credit institutions for larger farmers, whose total budget was one-third larger than the FSA's.[72] Therefore, while credit for larger farmers was apparently safe, credit for less substantial farmers was under attack.

Individual FSA clients provided complaints to receptive opponents of the agency. For example, in February 1942, in a letter to the Farm Bureau, J. H. Beam of Seneca, South Dakota, charged that FSA loans were loaded down with "big graft and red tape." Furthermore, he wrote, with grain and feed surpluses still mounting, government loans to produce more farm goods were unneeded.[73] Guy Earnest, a farmer from Ravenna, Nebraska, chimed in with his complaints as a rural rehabilitation client: "It would take me days to tell you all the dirty, rotten things that I have had to endure through the commands and directions of the local Farm Security Administration." Earnest concluded, "We hear about how bad it will be if Hitler takes us over, but the Farm Security Administration makes us wonder if this is really America or the tail end of Hitler's war machine?"[74]

The National Farm Bureau Federation was the most powerful opponent of the rural rehabilitation program. Family memberships in the Farm Bureau doubled from 21,000 to 42,000 in the Plains between 1940 and 1945. Kansas had the most member families with 32,000 in 1945. More significant was the farm organization's power on the national level. In 1940, 444,000 families had Farm Bureau memberships, and by the end of the war this number had increased to nearly 1 million families.[75] Therefore, during World War II the Bureau had the political and financial power, as well as the support from nonagricultural interests, to shape national farm policies. It was the national office, using pressure from farmers around the country, that persuaded Congress to kill the rural rehabilitation program, not the plains' state and local affiliates.

The Farm Bureau seems to have treated the rural rehabilitation program with benign neglect during the worst days of the Great Depression. Since the federal agency assisted even substantial farmers during emergency drought periods, it was not in the Bureau's interest to attack it or the New Deal. After 1936, however, the farm organization increasingly found fault with the USDA and the rural rehabilitation program. First, many in the Farm Bureau believed that the USDA ignored them, complaining that the Roosevelt administration's agricultural policy was set by federal officials, consumers, and organized labor.

Second, the Bureau had succeeded by great effort at building a symbiotic (some said parasitic) relationship between local Farm Bureaus and county agents. During World War II, the Bureau opposed the rural rehabilitation program primarily because the farm organization feared that a separate power base of county FSA offices in partnership with the Farmers Union would emerge in rural America. The FSA channeled funds into Farmers Union cooperatives, confirming this fear. Ironically, another influential farm group, the Grange, supported cuts in the FSA out of fear that another government agency–farmer alliance would arise like that of the Farm Bureau and the Agricultural Extension Service.

Finally, the Farm Bureau equated the interests of the FSA with those of organized labor. Perhaps the Bureau associated the FSA's leading supporter, the Farmers Union, with labor unions through their shared populist rhetoric and support for continuing the New Deal through World War II. Both the FSA and labor unions, in the Bureau's eyes, fostered federal intervention and potentially drove up labor costs. Therefore, the Farm Bureau increasingly joined with large business interests such as the Chamber of Commerce to criticize labor for promoting higher wages, shorter hours, and overtime pay, and for supporting the remaining New Deal programs. During the war, for example, the Bureau flexed its muscle to ensure its control over farm labor. The War Manpower Commission worked during the war to ensure adequate farm labor supplies and working conditions. In 1943 the Farm Bureau pressured Congress to move the War Manpower Commission into the Department of Agriculture. Within the USDA, the commission was prohibited from using federal funds to set housing standards, wages, and hours for farm workers, and was barred from promoting collective bargaining among farmers.

On the national level, the Bureau came to actively oppose organized labor. After the war, the organization's in-house historian noted that the Farm Bureau openly supported bills meant "to curb the excesses of unionized labor." The Bureau was especially alarmed when dairy farmers joined the United Mine Workers to form an autonomous local.[76] Would John L. Lewis organize smaller farmers as he had organized workers in the steel and rubber industries? Farm Bureau chapters in the plains were apparently concerned. The Bureau placed advertisements in local newspapers with the headline, "Why You Should Be a Farm Bureau Member—It's Organize or Be Organized." The Bureau warned against farmers joining unions sponsored by organized labor. "How unnatural it would be for farmers to follow such a course," since supposedly the unions did not understand farmers' problems.[77]

For these reasons, beginning in 1942 the national Farm Bureau office took an active role in attacking the FSA and its labor allies and in cutting funding for the rural rehabilitation program. The Bureau supplied "ammunition" to Senator Byrd's committee in its hearings on the FSA by sending investigators to states to uncover the agency's alleged misdeeds. Though the national Farm Bureau hurled many accusations at the FSA, its primary public arguments were that the FSA was a superfluous agency and its values ran counter to those of the countryside.

First, the Farm Bureau charged the FSA was truly a "nonessential" government program. In May 1943, *Nation's Agriculture,* the Bureau's official national magazine, charged that the FSA's supporters in labor unions and religious organizations were "under the delusion that all you have to do to increase farm production is to shovel out money to farmers." In fact, the real need was for farm labor and farm machinery. "Many of the submarginal farmers" that the FSA's supporters wanted to aid "are employed in war plants or on other farms."[78]

The second and most damning indictment the Farm Bureau used against the FSA was that it was fostering an alien economic philosophy. Bureau head Edward O'Neal charged that the tenant purchase program resembled "collectivist farming in Communist Russia."[79] In the last days of the FSA, the national Farm Bureau office delivered more blows to the agency's image. In addition to the FSA's alleged "maladministration" and "mismanagement," the FSA was allegedly attempting to "make America over." The leadership of the FSA had shown "a stealthy determination to do away with the individual ownership of farms, to collectivize agriculture and to regiment the producers of food under a collectivist system." Even though the FSA promised to liquidate its assets in farm purchase programs, *Nation's Agriculture* contended, the agency was attempting to impede its former clients with ten-year leases as a trick to continue its experiments in "collective farming."[80]

The spring of 1943 saw the fatal blow to the Farm Security Administration. Within the appropriation bill for that year, Congress cut funding for the rural rehabilitation program by more than half and severely limited the program's activities. President Roosevelt reluctantly signed the measure. Eventually, the agency perished on the vine through the steady funding constrictions. Congress passed the FSA's cases and projects onto the Farm Credit Administration, whose clientele consisted of the more substantial farm operators. By the end of World War II, Congress had slashed all funding for the rural rehabilitation program. However, commodity supports remained in place. Wartime expediency, economics, conservative ideology, and political pressures insured the demise of the first and the vitality of the second.

The New Deal and the rural rehabilitation program traveled a rocky road toward acceptance in the Great Plains states. The real threats of continued drought and descent into poverty drove plains farmers into the arms of President Roosevelt's farm program. Yet while the Plains farmers appreciated the assistance, many grew uneasy with its consequences. Farmers liked the government checks but not the accompanying restraints on expanding their production. The region's conservative politicians and press bemoaned that such assistance turned the farmer into a supplicant to the Washington bureaucrat. Worse for them was the prospect that the federal government would force the plowman in lockstep to its wishes. Furthermore, many of the region's important legislators and newspaper editors feared that bureaucrats from back East would turn the rich plains soil into a garden of experimental planning, or worse, take it out of production. Farmers themselves learned to cash the check and resent Washington for its aid. They justified this aid, since they believed AAA checks were supplements to their income, by likening it to the subsidies that industries had received for decades through tariffs.

Plains farmers and their leaders eventually accepted agricultural commodity subsidies begun under the AAA as a means to address long-term problems such as erratic farm income, huge grain surpluses, and soil erosion. In contrast, conservatives never saw rural rehabilitation as anything more than a short-term emergency measure, like federal work relief. While there was apparently no widespread opposition to rural rehabilitation in the plains, neither did the program have a committed and influential base of support in the region. The majority of plains voters, however, came to oppose the New Deal activism that the FSA embodied. Few agricultural representatives begrudged the smaller farm families for taking loans and grants to stay on their property during times of drought, low farm prices, and high unemployment during the 1930s, yet most assumed that once higher prices and rains returned, borderline farmers would have to sink or swim on their own again. The demise of the FSA and the fate of borderline farmers aroused resistance from religious groups, labor unions, and the Farmers Union. However, during the heady days of wartime prosperity, many in the plains farm sector, already frustrated by government intervention, allowed market forces and broker state politics to decide the fate of the rural rehabilitation program.

Conclusion

This study began with empty farmhouses in the plains and it ends there, too. During World War II renters deserted more and more farms. When farmers drove into the nearest town or village they saw notices in the stores and in the courthouse of those selling their livestock and implements and leaving farming permanently. Many could not keep up with the demands of wartime farming and were retiring. Others, who had spent the past decade trying to piece together an adequate income on the land, left the rural plains for military service or had found jobs in better-paying war industries. On the trip into town, when farmers sought out the rural rehabilitation farm supervisor in the courthouse, they found the office closed. By the middle of World War II, the Farm Security Administration was severely constricted through budget cuts. Its rural rehabilitation program was still collecting loan payments from clients, but most expected its operations to be liquidated soon.

Some plains observers yet again saw opportunity for aspiring farmers. Just as during the frontier days and World War I, many speculated that a rising farm economy would lead to another land boom in the plains countryside. *Capper's Farmer* surmised in 1943 that wartime economic jitters, confidence in postwar commodities markets, and low interest rates and land prices would lead investors to a rush back to the soil for a "safe" investment. Invariably, this would drive up land values.[1] The *Lincoln Star* also saw a vibrant economy fueling a land boom in Nebraska that year. Insurance companies and other holders of foreclosed land were finally selling it to tenants and private investors. Farm property was a secure investment again. The newspaper noted that "the number of Lincoln business and professional men who have purchased farms in the past year reads like a roster of the country club."[2]

Some saw the land boom and the improved commodity markets as signs of a reinvigorated plains countryside. The coming of spring 1943 stimulated the optimism of one newspaper in Williston, North Dakota, which celebrated the return of rural opportunity. The newspaper applauded the large harvests in recent years and predicted that undervalued land would lead veterans back home. "Soldiers in faraway lands are thinking of brown furrows across the Plains," the *Williams County Press* bragged. "Each surge in the wheat market, each raise in parity prices for agricultural products is another pull toward the soil." Compared to the inflation and food rationing that city-dwellers faced, the *Press* noted, the veteran could find "security, safety, health and family happiness, on a North Dakota farm."[3]

Even though FSA operations were ebbing away, some in the state saw the agency as the ideal vehicle for setting up farms for returning veterans in prosperous rural neighborhoods. Plains farmers who had served in the armed forces overseas were coming home to uncertain futures while deserted farm homes dotted the region. Perhaps, many reasoned, these veterans and their families could move in and start new farms to feed a world devastated by war. The lower house of the North Dakota legislature thought so, and it petitioned the U.S. Congress to issue rural rehabilitation and tenant purchase loans specifically for returning veterans. The legislators noted that "experience has demonstrated that loans made in the past under [the programs] have been highly satisfactory to the borrowers . . . and has afforded the greatest measure of security of home ownership as compared with any program of similar nature yet devised."[4] For this purpose, what was left of the FSA made 2,500 rural rehabilitation loans averaging $2,000 each, and 121 tenant purchase loans averaging $10,000 each, to veterans in the Great Plains states.[5]

In contrast to these sanguine outlooks, agricultural officials at the national and regional levels warned against looking to the farm as a refuge for nervous investors, a solution for postwar underemployment, or a home for discharged soldiers and sailors. Secretary of Agriculture Claude Wickard advised that highly productive mechanization adopted during the war had cut the need for more farmers. After wartime commodity markets subsided, he wrote, veterans could replace retiring farmers and women and youths working the fields. However, he cautioned, "if we are to maintain a good standard of living on the land, our farms should not be expected to support a greatly increased number of people after the war." Rural America, Wickard insisted, "should not be looked upon as a reservoir of opportunity" for the unemployed.[6]

Some within the USDA believed even the secretary's admonition was too optimistic. In 1944 a spokesman for the Farm Credit Administration noted that there were far more returning veterans desiring to start a farm than there were retiring farmers. Furthermore, the agency cautioned against the government making loans for prospective farmers with little capital.[7] The South Dakota extension service concurred, warning potential buyers not to purchase farmland without considering the consequences. It advised would-be farmers that agriculture was "still a highly competitive business . . . with relatively large costs that are hard to reduce, slow turnover, and narrow margin of profit." Looking back at three decades of land booms and busts, volatile commodity markets and environmental forces, and farm insolvency in the state, the extension service strongly advised against going into debt to purchase farmland at inflated prices.[8] These warnings from contemporary farm experts were quite a change from the boomer mentality that permeated the plains region during the frontier years and the 1920s. In contrast to celebrating the region's productive capacity and cheap land, state and federal officials by World War II had finally begun to warn people, especially undercapitalized borderline farmers, against making speculative farm investments in the region.

These warnings from federal and state officials were more realistic, based on recent history, than the cheery optimism of many in the press. After World War I, farmers in the Great Plains were caught up in the tide of increased production, overspeculation, and debt. While many farmers were able to ride the currents to prosperity, others, like the borderline farmers, were overwhelmed by the changes and sunk into poverty. Some borderline farmers sought refuge in subsistence farming or avoiding debt, strategies that offered them a haven from the rising tides of economic change. However, one could farm in place apart from powerful trends for only so long. Most either rose with the waves or left farming. Increased capital investment in the farm; greater mechanization, productivity, and acreage; involvement in government farm programs; and adapting to environmental conditions were the keys to staying afloat and succeeding during these volatile times that seemed to offer anything but security to plains farmers. However, borderline farmers and their families were out of their depth trying to compete during an era of expansion in the scale of plains agriculture. As a result, despite their attempts to attain middle-income security and status, these families lacked the means to transform themselves from "making do" to "having made it."

The New Deal agricultural programs offered critical assistance to farmers, but they worked best with those farmers hurt mainly by the Great Depression. The program did little for those who were already unstable before the Wall Street crash of 1929. Furthermore, the New Deal failed to take advantage of the reforming spirit of the time to restructure rural America. On one hand, New Dealers found strong support for federal intervention through agricultural commodity supports, mortgage refinancing, and soil conservation programs when such efforts matched the goals of plains farmers. Eventually, even Republicans in the region accepted these programs, and they worked to make them USDA, rather than Democratic, initiatives.

However, unlike the more popular programs, rural rehabilitation never progressed beyond its status as an "emergency" initiative. Like the Works Progress Administration, it failed to establish itself as a permanent program. The rural rehabilitation program took on clients between 1935 and 1942, but before 1939 it acted primarily as a rural relief agency. Afterward, its main purpose was to rehabilitate the most promising borderline farmers. However, four years was not enough time to conclusively measure the progress of its clients or the program.

The rural rehabilitation program succeeded, through small grants and loans and advice, in improving these borderline farms. In the tenant purchase program, the federal government transformed select farm renters into owners, often with income comparable to the region's general farmers. The RA and FSA found thousands of insolvent farm families with diminished standards of living on deteriorating farms with deteriorating soils. These operations were undersized, often unadapted to the plains environment, overly committed to commodity markets, and poorly managed. The rural rehabilitation program stepped in to make the operators more secure in diet, soil fertility, tenure, and farm management. As a result, the activist government initiative was a boon to those plains farm families on the margins of poverty during the thirties. Perhaps most important, the greatest triumph of the Resettlement Administration and the Farm Security Administration was in showing borderline farmers around the country that they mattered and that their government was concerned about them in the face of the worst economic depression in American history. However, there were limitations to the program. Confronting vast economic, agricultural, social, and environmental forces, rural rehabilitation was underfunded, understaffed, and lacked the support to truly correct the problems that troubled borderline plains farmers.

The stories of the RA and FSA reveal the limitations of the New Deal. British historian Anthony Badger believes the New Deal agencies constructed a "holding pattern" that maintained the social and economic status quo until economic recovery. The United States government lacked, through the will of its citizens, the "state capacity" to truly reform America. Through the emerging "broker state," the New Deal constructed its own support system of bureaucrats, lobbyists, and influential congressional committee members. The rural rehabilitation program acted as a "triage" to stabilize the more promising borderline farmers until commodity prices rose once more. The New Deal farm program brought unprecedented government intervention into the countryside. However, the nation had neither the bureaucracy, the funding, nor the will to reform rural America actively and decisively, as the tenant purchase program potentially could have done. Many plains Republicans, for example, made the leap from Herbert Hoover's limited state to Franklin Roosevelt's welfare state for market-oriented farmers, but they drew the line at Rexford Tugwell's plans to use the federal bureaucracy to reform rural America.

Finally, wartime prosperity and political pressures bore down to constrict FSA funding, and broker-state machinations pulled the rug out from underneath rural rehabilitation. The program had the impassioned support of most Farmers Union members. Religious and labor organizations, which had become a vital part of the New Deal's urban coalition, also came forward to support the FSA. However, most voters wanted the economy, rather than the government, to shape the countryside. This left the fate of rural rehabilitation in the hands of the power brokers. The preeminent forces in setting American agricultural policy, the national Farm Bureau and powerful congressional members of agricultural committees, indicated that the time for the rural rehabilitation program had passed.

The story of the RA and FSA is important not only to the hundreds of thousands of farm families the agencies assisted. The demise of the rural rehabilitation program foretold the fate of rural America. Rather than embracing a countryside populated with smaller and medium-scale farm operations, Americans espoused large-scale farms that only upper-income families could afford to purchase and run profitably. This made this country's farms a marvel of productivity. However, this choice betrayed the democratic ideals that family-based agriculture was supposed to represent and that the nation's founders and the proponents of the Homestead Act had celebrated. Americans decided that farming was a business, not a way of life. However, especially in the semiarid plains, the dependency on fuel-based tractor power and fertilizers called into question whether modern farming methods could be indefinitely sustained both for economic and environmental reasons. One critic of modern commercial agriculture castigates this "fossil-fuel chemotherapy" for giving us "a false sense of the health of the agricultural system."9 The conflict of modern plains farming continues to pit volatile economic, environmental, and agricultural forces against a stable way of life.

Furthermore, the story of the plains rural rehabilitation tells us a great deal about Americans' relationship with their government. Decades ago, Arthur Schlesinger Jr. portrayed the reforms of New Deal as part of a cyclical advance in America, leading to a progressively greater involvement of government in daily life. However, the politically conservative Reagan-Bush era of the 1980s was followed by the fiscally conservative Clinton presidency of the 1990s. This in turn was followed by another politically conservative Bush administration in the new century. During this time, political trends seem to challenge New Deal–style intervention in American life. On one hand, Social Security pensions and Medicare health care for the elderly appear exempt from cutbacks. A huge, broad-based, and well-organized "deserving" constituency supports both programs. Both have the public endorsement as a type of "insurance" funded by employee/employer payroll taxes. But even their supporters admit critical flaws, since both programs are funded by regressive payroll taxes and require a huge portion of the federal budget. However, because of their reputations as limited, semiprivately funded programs, Social Security and Medicare have risen to the pantheon of "entitlements." That is, their recipients believe they are qualified to benefits from the two programs once they have reached old age.

In contrast, government programs that appear to constrict opportunity or that assist the "undeserving poor" met the fiscal chopping block. Americans have a historic suspicion of the federal government. As one writer notes, for them "government is accepted as, at best, a necessary evil, one we must put up with while resenting the necessity." This belief, according to Garry Wills, "belittles America [for it] asks us to love our country by hating our government."[10] This mistrust also leaves millions without basic services such as medical insurance. During the mid-1990s the nation's leading conservative Republican stated that bureaucracies "are inherently due to fail in a world that is too complex and too human to be limited and rational." Modern liberals, stated Speaker of the House Newt Gingrich, "have forgotten that government cannot substitute for personal responsibility, or faith." Democratic president Bill Clinton responded, not with a passionate defense of government intervention, but with an admission that "the American people are torn about what role government ought to play. They say they can't stand big government and they want less of it—but they have huge aspirations for it." He also conceded that in recent years, many Americans were "angry and frustrated because prosperity and stability do not cover all who work hard and play by the rules, and because they feel that government is helping special interests, and not holding everyone equally accountable."[11]

In the conservative, antigovernment spirit of the times, the Republican-dominated United States Congress, with the concurrence of the Democratic president, carved off important government agricultural and social programs. In 1996 Congress passed unprecedented changes in well-established agricultural commodities and social welfare programs set in place by New Dealers six decades before. That year, Congress under the Federal Agricultural Improvement and Reform (FAIR) Act, also known as the Freedom to Farm Act, set about to end New Deal agricultural commodity payments. Under the law the federal government compensated farmers with payments over seven years, protected them against low market prices, and provided them payments for retiring land into a conservation program. The government also maintained a four-million-ton grain reserve to support prices. As planned, after seven years FAIR would withdraw the government from its roles as a buyer, storehouse, and stockholder of American grains and fibers. At that point, supporters predicted, farmers would profit from commodity markets, free from government constraints. Some critics, however, doubted this and have called FAIR the "Freedom to Fail" farm act. Events proved them right when Congress reinstated expensive farm payments when agricultural commodity prices fell in the years after FAIR's passage.

On the other hand, Congress killed another New Deal–era program and kept it dead. In 1996 Congress also abolished Aid to Families with Dependent Children (AFDC, popularly known as welfare). In an effort to "end welfare as we know it," Congress handed the program over to the states with certain provisions to drop the "undeserving poor" and pressure the "deserving poor" to find work or face being cut from welfare. Like the borderline farm families a half century before them, these families had to fend for themselves in the changing, often tempestuous economy.

This story of empty farmhouses in the Great Plains, and the federal government's programs to assist their former residents, has significant regional and national implications. After the Great Depression, Americans and their government consciously chose not to maintain a large population of comparatively small, relatively inefficient, capital-starved farms on the land. This left government commodities programs, market forces, and the increasing scale of agriculture alone to shape the plains country-side. By 1997 the average plains farm had increased to over one thousand acres in size. It was valued at one-half million dollars, with an average debt of $132,000. Between 1990 and 1997 twenty thousand farms in the region disappeared.[12] Farms that were larger, more expensive, and more productive probably absorbed these operations.

This leads one to wonder about the future of farm life in the plains, and whether it will remain a way of life that represents the best this country has to offer. Although many of these remaining farms provide a comfortable lifestyle equal to one in the city, one wonders whether that is enough. During World War II, Earl Bell warned that for many, "farming as a mode of life was sacrificed in favor of what they considered a fuller life and one that could be bought."[13] Has the push for maximum production betrayed the real importance of the family farm in American life? Shouldn't farming offer more than a way to acquire more goods through greater yields? A familiar refrain in modern American agriculture celebrates exploding productivity since 1940. Yet, as one writer states, boasts about increased yields "are in truth concessions that farming offers the public nothing more than meat and potatoes."[14] Farm life as Americans have known it is dead. Only the future knows what it will become.

NOTES

Introduction

1. Department of the Interior, National Park Service, *Historical Overview and Inventory of the Niobrara/Missouri National Scenic Riverways, Nebraska/South Dakota*, by Rachel Franklin, Michael J. Grant, and Martha Hunt (Omaha, Nebr.: National Park Service, 1994), 36–37.
2. Bureau of the Census, *Fifteenth Census of the United States: 1930, Agriculture*, vol. 3: *Type of Farm*, pt. 1: *The Northern States* (Washington DC: GPO, 1932), county table 1; Bureau of the Census, *Sixteenth Census of the United States: 1940, Agriculture*, vol. 2, *Third Series: State Reports*, pt. 1: *Statistics for Counties* (Washington DC: GPO, 1942), county table 19; Bureau of the Census, *United States Census of Agriculture: 1945, Statistics for Counties*, vol. 1, pts. 11–13 (Washington DC: GPO, 1946), state table 8.
3. Department of Commerce, *Historical Statistics of the United States: Colonial Times to 1970*, pt. 1 (Washington DC: GPO, 1975), 457.
4. United Nations, Food and Agricultural Organization, *Production Yearbook, 1962* (n.p.), 18.

1. Borderline Farmers

1. Lorena Hickok to Anna Eleanor Roosevelt, Nov. 1933, Folder: Excerpts from Lorena Hickok's Letters to Eleanor Roosevelt and Her Reports to Henry Hopkins, Sept.–Dec. 1933, North and South Dakota, Nebraska, Iowa, Minnesota, and New York City, Federal Emergency Relief Administration Papers (hereafter FERA Papers), Franklin D. Roosevelt Library (hereafter FDRL), Hyde Park, N.Y.
2. Lorena A. Hickok to Harry Hopkins, Dickinson, N.Dak., Oct. 30, 1933, Folder: North Dakota, Field Reports 1933–1936, FERA Papers, FDRL.
3. Department of Commerce, *Historical Statistics*, 461.
4. Department of Commerce, *Historical Statistics*, 511.
5. Department of Commerce, *Historical Statistics*, 500.
6. Department of Commerce, *Statistical Abstract, 1932* (Washington DC: GPO, 1933), 303, 593; Department of Commerce, *Statistical Abstract, 1941* (Washington DC: GPO, 1942), 680, 701.
7. Bureau of the Census, *United States Census of Agriculture: 1945*, vol. 1, pts. 11–13, state table 6.

8. David Danbom, *Born in the Country: A History of Rural America* (Baltimore: Johns Hopkins University Press, 1995), 194.

9. Val Kuska, Burlington Railroad Colonization Agent, to Ralph Budd, Omaha, Nebr., Apr. 11, 1933, File 77-A-1, Misc. Correspondence, Clippings 1925–1940, Series: Agricultural Situation, Val Kuska Papers, MS1431 (hereafter Kuska Papers), Manuscripts Division, Nebraska State Historical Society, Lincoln (hereafter NSHS).

10. Bureau of the Census, *Sixteenth Census of the United States: 1940, Agriculture*, vol. 1, *First and Second Series: State Reports*, pt. 2: *Statistics for Counties* (Washington DC: GPO, 1942), county table 8.

11. Department of Agriculture, *The Farm Real Estate Situation, 1933–34*, Circular no. 354 (Washington DC: GPO, April 1935), 30.

12. Lee J. Alston, "Farm Foreclosures in the United States during the Interwar Period," *Journal of Economic History* 43 (1983): 888.

13. Lawrence A. Jones and David Durand, *Mortgage Lending Experience in Agriculture* (Princeton: Princeton University Press, 1954), 68–73, 28.

14. South Dakota, Agricultural Experiment Station, *Indebtedness on 48 Potter County Farms, 1930*, Circular no. 2, (Brookings: South Dakota Agricultural Experiment Station, Mar. 1932), 3–4.

15. South Dakota, Agricultural Experiment Station, *Indebtedness on 48 Potter County Farms, 1930*, 7.

16. H. S. Ewen to Franklin D. Roosevelt, Carrington, N.Dak., Oct. 29, 1932, Folder: North Dakota After Elections—E–G, DNC Campaign Correspondence, 1928– 1933, Democratic National Committee Papers (hereafter DNC Papers), FDRL.

17. T. F. Green, Cashier, to L. A. White, Rehabilitation Director, Valley, Nebr., Sept. 25, 1936, File 209, Resettlement Administration, Series 1, Correspondence of the Office of the Governor, 1935–1940, Box 13, Governor Robert Cochran Papers, Archives Division, NSHS.

18. William L. Bates, "Farm Loan Commissions," *Nebraska Farmer*, Nov. 26, 1932, 7.

19. Resettlement Administration, Region VII, Land Use Planning Section, Land Utilization Division, "Background Data of Region VII," mimeo (Lincoln, Nebr.: Resettlement Administration, July 1937), 30, 59, 87, 42, 56, 87, 71.

20. Brian Q. Cannon, "Immigrants in American Agriculture," *Agricultural History* 65 (1991): 19, 30; Kathleen Conzen, "Peasant Pioneers: Generational Succession among German Farmers in Frontier Minnesota," in *The Countryside in the Age of Capitalist Transformation: Essays in the Social History of Rural America*, ed. Steven Hahn and Jonathan Prude (Chapel Hill: University of North Carolina Press, 1985), 275.

21. Kansas, Agricultural Experiment Station, *Drought in Kansas*, Bulletin no. 547 (Manhattan, Kans., 1971), 3.

22. Works Progress Administration, *Areas of Intense Drought Distress 1930–1936*, Division of Social Research Series V, no. 1 (Washington DC: GPO, Jan. 1937), 11.

23. Kansas, Agricultural Experiment Station, *Drought in Kansas*, 6–7.

24. "Agricultural Conditions in Santa Fe States, July 1, 1934," Crop Reports Folder RR394.4, Atchison, Topeka, & Santa Fe Railway Co., Agricultural Development and Publicity Office, Kansas State Historical Society, Topeka (hereafter KSHS).

25. J. O. Shroyer to Val Kuska, Humboldt, Nebr., Aug. 3, 1931, File 78-G-1, Drought Relief, Aug. 1931–Oct. 1932, Series: Agricultural Relief Programs, Kuska Papers, NSHS.

26. John R. Snyder, Chairman, Box Butte County Board, to Governor R. L. Cochran, Alliance, Nebr., File no. 321, Cochran Papers, NSHS.

27. "Barren Soil Succumbs to Drouth, Wind," *Denver Post,* May 5, 1935.

28. "Fears Worse Dust Storms," *Omaha World-Herald,* Sept. 20, 1936.

29. South Dakota, Department of Agriculture, *Annual Report, 1933–1934* (n.p.), 9.

30. Wolff, Gerald W., and Joseph H. Cash, compilers and editors, "South Dakotans Remember the Great Depression," *South Dakota History* 19 (1989): 230.

31. "Millions of Migratory Grasshoppers Swarm across the Skies of Nebraska," *Lincoln Star,* June 24, 1936.

32. Lorena Hickok to Anna Eleanor Roosevelt, Huron, S.Dak., Nov. 11–12, 1933, FERA Papers, FDRL.

33. Wolff and Cash, "South Dakotans Remember the Great Depression," 236, 231.

34. Harold Bennet Clingerman, *Field Man: The Chronicle of a Bank Farm Manager in the 1940s* (Ames: Iowa State University Press, 1989), 43–44.

35. South Dakota, Agricultural Experiment Station, *Estimated Returns from Farms of Large, Medium, and Small Size of Business in the Spring Wheat Area of South Dakota,* Circular no. 20 (Brookings: South Dakota Agricultural Experiment Station, May 1934), 3–13.

36. South Dakota, Agricultural Extension Service, *Indebtedness on 48 Potter County Farms, 1930,* 11; South Dakota, Agricultural Extension Service, *Estimated Returns from Farms,* 23.

37. Richard Bremer, *Agricultural Change in an Urban Age: The Loup Country of Nebraska, 1910–1970,* University of Nebraska Studies, new series no. 51 (Lincoln: University of Nebraska, June 1976), 59, 147.

38. Bremer, *Agricultural Change in an Urban Age,* 65–66.

39. Hargreaves, Mary W. M., *Dry Farming in the Northern Great Plains: Years of Readjustment, 1920–1990* (Lawrence: University Press of Kansas, 1993), 77.

40. Gordon Morris Bakken, *Surviving the North Dakota Depression* (Pasadena, Calif.: Wood and Jones, 1992), 7, 14.

41. Deborah Fink, *Agrarian Women: Wives and Mothers in Rural Nebraska, 1880–1940* (Chapel Hill: University of North Carolina Press, 1992), 102–4.

42. Hargreaves, *Dry Farming in the Northern Great Plains: Years of Readjustment, 1920–1990,* 64.

43. Bureau of the Census, *United States Census of Agriculture: 1945,* vol. 1, pts. 11–13, state table 1.

44. David B. Danbom, "The North Dakota Agricultural Experiment Station and the Struggle to Create a Dairy State," *Agricultural History* 63 (1989): 178–80.

45. Department of Agriculture, Bureau of Agricultural Economics, *Culture of a Contemporary Rural Community: Sublette, Kansas,* by Earl Bell, Rural Life Studies no. 2 (Washington DC: GPO, 1942), 25, 42.

46. Brookings study cited in James T. Patterson, *America's Struggle against Poverty 1900–1980* (Cambridge: Harvard University Press, 1981), 16; South Dakota figures from South Dakota, Agricultural Experiment Station, *A Graphic Summary of the Relief*

Situation in South Dakota (1930–1935), Bulletin no. 310 (Brookings: South Dakota, Agricultural Experiment Station, May 1937), 31.

47. Bureau of the Census, *United States Census of Agriculture, 1945*, vol. 1, pts. 11–13, state table 7.

48. Bureau of the Census, *United States Census of Agriculture: 1945*, vol. 2: *General Report, Statistics by Subject* (Washington DC: GPO, 1947), 162, 602.

49. Bureau of the Census, *Sixteenth Census: 1940, Agriculture*, vol. 2, pt. 1, county table 17; Department of Commerce, *Historical Statistics*, 459, 483.

50. Department of Commerce, *Historical Statistics*, 126, 167.

51. From map in Bureau of the Census, *Sixteenth Census: 1940, Agriculture*, vol. 3: *General Report, Statistics by Subject* (Washington DC: GPO, 1943), 875.

52. Bureau of the Census, *Sixteenth Census: 1940: Agriculture*, 3:875.

53. Department of Agriculture, *Family Income and Expenditures: Pacific Region and Plains and Mountain Region*, pt. 1: *Family Income*, USDA Miscellaneous Publication no. 356 (Washington DC: GPO, 1939), 178–79; Department of Commerce, *Historical Statistics*, 461.

54. Department of Commerce, *Historical Statistics*, 463.

55. Department of Agriculture, *Family Income and Expenditures*, 201–2, 214–15.

56. Department of Commerce and Department of Agriculture, *Analysis of Specified Farm Characteristics for Farms Classified by Total Value of Products* (Washington DC: GPO, 1943), table 4.

57. Bureau of the Census, *Sixteenth Census: 1940, Agriculture*, vol. 1, pt. 2, state table 9.

58. Department of Agriculture, *Family Income and Expenditures*, 193. Figures rounded.

59. Department of Agriculture, *Family Income and Expenditures*, 178–79.

60. Bureau of the Census, *Sixteenth Census: 1940, Agriculture*, vol. 2, pt. 1, county table 20.

61. Department of Agriculture, *Family Income and Expenditures*, 201–2.

62. O. J. Maurer to Franklin D. Roosevelt, Pukwana, S.Dak., Dec. 16, 1932, Folder: South Dakota after the Election—M, DNC Campaign Correspondence, 1928–1933, DNC Papers, FDRL.

2. Farm Tenancy in the Great Plains

1. "Like a Cry from the Wilderness," *Wichita Independent*, Folder: General Newspapers and Magazine Articles, General Correspondence Maintained, Washington Office, 1935–38, Records of the Farm Home Administration, Record Group 96 (hereafter FHA Records, RG 96), National Archives, College Park, Md.

2. South Dakota, State Planning Board, *Land Ownership in South Dakota* (Brookings: State Planning Board, 1937), 3.

3. Washington to William Strickland, Mount Vernon, Va., July 15, 1797, from Wayne D. Rasmussen, ed., *Agriculture in the United States: A Documentary History* (New York: Random House, 1975), 1:290–91.

4. Department of Agriculture, Bureau of Agricultural Economics, *Graphic Summarization of Farm Tenure Based on 1940 Census* (Washington DC, 1946), table 1a.

5. Bureau of Agricultural Economics, *Graphic Summarization of Farm Tenure Based on 1940 Census*, fig. 4.

6. Benson Y. Landis, *Sedgwick County, Kansas: A Church and Community Survey* (New York: George H. Doran, 1922), 76.

7. Harry R. O'Brien, "Tenancy At Its Worst," *Country Gentleman*, Jan. 4, 1919, 3.

8. Henry J. Allen, "Smoking Out the Land Hogs," *Country Gentleman*, Dec. 6, 1919, 3.

9. William Johnson, "Rent or Go to Town," *Country Gentleman*, Dec. 11, 1920, 15, 50.

10. "Jacob, Henry, William—and Frank," *Breeder's Gazette*, Feb. 19, 1920, 455.

11. Bert B. McConachie to Congressman Clifford Hope, June 5, 1930, Folder: Dept. Corres., 1930 Agriculture, Corporation Farming, Gen. Corres. 1930 and Dept. Corres. 1930, Agriculture to Post Office, Clifford Hope Papers, KSHS.

12. Rev. Leonard H. Torline to Franklin D. Roosevelt, Dubuque, Kans., Oct. 11, 1935, Folder: Kansas, Clergy Letters, President's Personal File no. 21A (hereafter PPF), FDRL.

13. L. R. French to Clifford Hope, Pretty Prairie, Kans., Jan. 6, 1937, Folder: Farm Tenant (Legis.), Legis. Correspondence, 1936–1937, General Correspondence, Hope Papers, KSHS.

14. Frank Bean to Clifford Hope, St. John, Kans., Aug. 7, 1938, Folder: Dept. Corres., 1938–1939, Farm Tenant, General Correspondence, Hope Papers, KSHS.

15. Bradley Baltensperger, "Farm Consolidation in the Northern and Central States of the Great Plains," *Great Plains Quarterly* 7 (1987): 258–61.

16. Thomas J. Pressly and William H. Scofield, *Real Estate Values in the United States by Counties, 1850–1959* (Seattle: University of Washington Press, 1965), 38–42.

17. Thomas Leadley, "Tenancy: Deep-rooted Disease," *Nebraska Farmer*, Jan. 30, 1937, 5.

18. William G. Murray, *Farm Appraisal and Valuation*, 5th ed. (Ames: Iowa State University Press, 1969), 507.

19. Robert Diller, *Farm Ownership, Tenancy, and Land Use in a Nebraska Community* (Chicago: University of Chicago Press, 1941), 37–53.

20. Diller, *Farm Ownership in a Nebraska Community*, 37–53.

21. Bureau of Agricultural Economics, *Graphic Summarization of Farm Tenure Based on 1940 Census*, table 9a.

22. Nebraska, Agricultural Experiment Station, *Farm Tenancy in Box Butte County, Nebraska*, Bulletin no. 336 (Lincoln: Nebraska Agricultural Experiment Station, Jan. 1942), 13–15.

23. Nebraska, Agricultural Experiment Station, *Farm Tenancy in Box Butte County*, 15–18.

24. Nebraska, Agricultural Experiment Station, *Farm Tenancy in Box Butte County*, 13–18.

25. James Malin, *The Grassland of North America* (Gloucester, Mass., 1967), 314.

26. "The Reflections of a Tenant," *Country Gentleman*, May 27, 1916, 1104.

27. Department of Agriculture, *The Ownership of Tenant Farms in the North Central States*, Department Bulletin no. 1433 (Washington DC: GPO, Sept. 1926), 38.

28. J. H. Beckmann to Franklin D. Roosevelt, Albion, Nebr., Oct. 23, 1935, Folder: Nebraska Letters, Clergy Letters, PPF21A, FDRL.

29. "Rural Resettlement Program Under Way," *Colton (S.Dak.) Courier*, Oct. 3, 1935.

30. William H. Harbaugh, "Twentieth-Century Tenancy and Soil Conservation: Some Comparisons and Questions," *Agricultural History* 66 (1992): 97, 98, 102.

31. Department of Agriculture, *Family Income and Expenditures*, 178–79.

32. South Dakota, Agricultural Experiment Station, *The Standard of Living of Farm and Vil-*

lage Families in Six South Dakota Counties, 1935, Bulletin no. 320 (Brookings: South Dakota Agricultural Experiment Station, Mar. 1938), 39.

33. Department of Agriculture, Family Income and Expenditures, 178–79.

34. Bureau of Agricultural Economics, Graphic Summarization of Farm Tenure Based on 1940 Census, fig. 5.

35. Petition, Folder: Legislative Correspondence, 1935–1936, Farm Relief: General, Hope Papers, KSHS.

36. Resettlement Administration, Compensation as a Means of Improving the Farm Tenancy System, Land-Use Planning Publication no. 14 (Washington DC, Feb. 1937), 86.

37. Diller, Farm Ownership in a Nebraska Community, 59–61.

38. Clingerman, Field Man, 78.

39. Resettlement Administration, "Background Data of Region VII," 48.

40. Department of Agriculture, Family Income and Expenditures, 178–79.

41. Diller, Farm Ownership in a Nebraska Community, 66.

42. Clingerman, Field Man, 35, 74, 114–15.

43. R. H. Burch to Clifford Hope, Haviland, Kans., Jan. 25, 1938, Folder: Farm Tenant Legis., Hope Papers, KSHS.

44. Fink, Agrarian Women, 98–102.

45. James Marten, "Continuity and Change on the Twentieth-Century Farm: The Gists of South Dakota, 1921–1971," Great Plains Quarterly 11 (1991): 39, 43–45, 48.

46. Jeremy Atack, "The Agricultural Ladder Revisited: A New Look at an Old Question with Some Data for 1860," Agricultural History 63 (1989): 14–15, 20–23.

47. Allan Bogue, "Foreclosure Tenancy on the Northern Great Plains," Agricultural History 39 (1965): 22.

48. "Tenants Buy Many Farms, Officials Told," Omaha World-Herald, Dec. 5, 1936.

49. Works Progress Administration, Relief and Rehabilitation in the Drought Area, Division of Social Research Series, vol. 3 (Washington DC: GPO, June 1937), 50–51.

50. Beckmann to Roosevelt, Oct. 23, 1935, FDRL.

51. "More Tenant Farms," editorial, Lincoln Star, Oct. 14, 1936.

52. W. H. Brokaw, "Farm Tenancy," speech, File 321, Cochran Papers, NSHS.

53. Congressional Record, 75th Cong., 1st sess., July 1, 1937, 6670.

54. Henry A. Wallace, "Rural Poverty, Remarks at the General Assembly of the Council of State Governments," Jan. 23, 1937, USDA Press Release 1060-37, from Rasmussen, Agriculture in the United States: A Documentary History, 2:2846.

55. Robert S. McElvaine, The Great Depression: America, 1929–1941 (New York: Times Books, 1993), 220–21.

56. Chamber of Commerce of the United States, Farm Tenancy in the United States (Washington DC: Chamber of Commerce, 1937), 24–29.

57. Charles Morrow Wilson, The Landscape of Rural Poverty: Corn Bread and Creek Water (New York: Henry Holt, 1940), 57.

58. Ray Yarnell, "Between Thee and Me," editorial, Capper's Farmer, Jan. 1937, 52.

59. Diller, Farm Ownership in a Nebraska Community, 55–57.

3. The Development of Rural Rehabilitation

1. M. W. Lusk to Herbert Hoover, Yankton, S.Dak., June 18, 1931, Folder: Farm Matters: Correspondence, 1931, Presidential Papers Subject File, Herbert Hoover Presidential Library (hereafter HHPL).

2. Quoted in McElvaine, *The Great Depression,* 55–56.

3. Department of Commerce, *Statistical Abstract of the United States, 1944–1945* (Washington DC: GPO, 1945), 416.

4. James C. Duram, "Constitutional Conservatism: The Kansas Press and the New Deal Era as a Case Study," *Kansas Historical Quarterly* 43 (1977): 447.

5. "Shall the Government Run Farming?" editorial, *Kansas City Star,* June 8, 1928.

6. Department of Agriculture, Federal Farm Board, *Third Annual Report of the Federal Farm Board for the Year Ending June 30, 1932* (Washington DC: GPO, 1932), 110.

7. "Our Agricultural Policy," editorial, *Omaha Daily Journal Stockman,* May 4, 1930.

8. South Dakota, Agricultural Experiment Station, *Summary of the Relief Situation,* 45.

9. McElvaine, *The Great Depression,* 61.

10. "Loans to Farmers Acts of 'Prize Boob,'" *Omaha World-Herald,* Mar. 30, 1932.

11. "Charity, Unemployment and Laziness," editorial, *Aberdeen (S.Dak.) News,* Oct. 15, 1930.

12. "Relief is First a Local Problem," editorial, *Kansas City Star,* Dec. 31, 1931.

13. F. W. Morse to Herbert Hoover, Gas City, Kans., Aug. 28, 1932, Folder: Farm Matters: Correspondence, Jan.–Sept. 1932, Presidential Papers Subject File, HHPL.

14. Theodore Saloutos, *The American Farmer and the New Deal* (Ames: Iowa State University Press, 1982), 270.

15. McElvaine, *The Great Depression,* xxv, xxiv.

16. McElvaine, *The Great Depression,* 91–92.

17. Mrs. Estella M. Beems to Franklin D. Roosevelt, Mills, Nebr., n.d., Folder: Nebraska After Election—B, Campaign Correspondence, 1928–33, DNC Papers, FDRL.

18. "Loans to Farmers," *Omaha World-Herald.*

19. "Text of President's Talk Given at Devils Lake Tues., Aug. 7," *Bowbells (N.Dak.) Tribune,* Aug. 7, 1934.

20. Charles and Joyce Conrad, *50 Years: North Dakota Farmers Union* (n.p., 1976), 54.

21. Elizabeth Evenson Williams, *Emil Loriks, Builder of a New Economic Order* (Sioux Falls, S.Dak.: Center for Western Studies, 1987), 57, 82.

22. "No Mandate for Ordered System," editorial, *Nebraska Union Farmer,* Nov. 25, 1936, 4.

23. Williams, *Emil Loriks,* 57.

24. Drouth Conference of the Midwest Farmers Union State Organizations to President Roosevelt, et al., Aberdeen, S.Dak., July 6, 1937, Folder: Official File (hereafter OF) 987, FDRL.

25. M. W. Thatcher, Farmers National Grain Corporation, to Congress, Chevy Chase, Md., Jan. 30, 1934, Folder: Agriculture—Misc. Papers, Dept. Correspondence, 1934, General Correspondence, Hope Papers, KSHS.

26. Tom Berry to Harry Hopkins, Pierre, S.Dak., Folder: Autumn 1934, State of the Nation Report, Harry L. Hopkins Papers, FDRL.

27. Works Progress Administration, *Final Statistical Report of the* FERA (Washington DC: GPO, 1942), 137–41, 263–64, 274–75, 279–80, 285–86.

28. "Work or Dole?" *Lincoln Star,* Feb. 7, 1935.

29. Federal Emergency Relief Administration, "Monthly Earnings of Rural Relief Households at the Time of Closing Cases," Research Bulletin Series II, no. 4 (June 11, 1935), mimeo, 4.

30. Nebraska, Emergency Relief Administration, *Work Relief in Nebraska, April 1934–July 1935* (Lincoln, Nebr.: Emergency Relief Administration, Work Division, Nov. 1935), 1.

31. South Dakota, Agricultural Experiment Station, and Federal Emergency Relief Administration, *Rural Relief in South Dakota,* Bulletin no. 289 (Brookings: South Dakota Agricultural Experiment Station, June 1934), 52.

32. "Relief Crackdown," *Business Week,* July 27, 1935, 14–15.

33. Works Progress Administration, "Survey of Rural Relief Cases Closed for Administrative Reasons in South Dakota," Research Bulletin Series II, no. 12 (Jan. 23, 1936), mimeo, 3, 7–8.

34. Works Progress Administration, *Survey of Workers Separated from* WPA *Employment in Nine Areas, 1937* (Washington DC: GPO, 1938), 12, 14, 15.

35. "Dust Relief Hits Pride," *Kansas City Star,* April 4, 1935.

36. Julia L. Miller, "Conditions in Western Kansas," RH MS 327:3:48, John Stutz Papers, Kansas Collection, University of Kansas, Lawrence.

37. Quoted from Pamela Riney-Kehrberg, *Rooted in Dust: Surviving Drought and Depression in Southwestern Kansas* (Lawrence: University Press of Kansas, 1994), 87.

38. "Farmers May Get WPA Work Without 'Relief' Registration, Sen. Nye Asserts," *North Dakota Nonpartisan,* Aug. 12, 1936.

39. Wolff and Cash, "South Dakotans Remember the Great Depression," 257–58.

40. Wolff and Cash, "South Dakotans Remember the Great Depression," 228–29.

41. "As to Direct Relief," editorial, *Lincoln Star,* May 29, 1936.

42. Works Progress Administration, "Relief and Rehabilitation in the Drought Area," *Division of Social Research Series,* vol. 3, (Washington DC: GPO, 1937), 3, 23.

43. Farm Credit Administration, "1936 Emergency Crop and Feed Loans," Jan. 9, 1937, mimeo.

44. Ralph Perkins, "Relief Work in a Dust Bowl County," *Sociology and Social Research* 23 (1939): 541.

45. Works Progress Administration, "Relief and Rehabilitation in the Drought Area," 4, 51.

46. *Bartlett's Familiar Quotations,* 15th ed. (Boston: Little, Brown, 1980), 779.

47. William Allen White to Herbert Hoover, Emporia, Kans., May 3, 1934, Folder: William Allen White Correspondence, 1933–35, Individuals File Series, Post-Presidential Papers, HHPL.

48. Hargreaves, *Dry Farming in the Northern Great Plains: Years of Readjustment, 1920–1990,* 60–64.

49. Richard S. Kirkendall, *Social Scientists and Farm Politics in the Age of Roosevelt* (Columbia: University of Missouri Press, 1966), 58, 84.

50. Cliff Hope Jr., *Quiet Courage: Kansas Congressman Clifford R. Hope* (Manhattan, Kans.: Sunflower University Press, 1997), 116.

51. Hargreaves, *Dry Farming in the Northern Great Plains: Years of Readjustment, 1920–1990*, 103, 109, 124.
52. "Submarginal," editorial, *Rapid City (S.Dak.) Journal*, Aug. 15, 1935.
53. E. R. Jonson to Sherman Johnson, Regional Director, Land Planning Program, Resettlement Administration, Rockford, Ill., May 7, 1935, unprocessed, J. Franklin Thackery Papers, HHPL.
54. Mrs. Philip W. Towsley to Sherman Johnson, Regional Director, Land Planning Program, Resettlement Administration, Springfield, Vt., n.d., unprocessed, Thackery Papers, HHPL.
55. Resettlement Administration, *First Annual Report of the Resettlement Administration: 1936* (Washington DC: GPO, 1936), 135.
56. "591,681 Allocated for West N.D. Land Project," *Fargo Forum*, Oct. 24, 1935.
57. "Purchase of Whole Town by Government Is Asked," *New York Times*, Dec. 29, 1938.
58. "Kearney Man Tells Governor of His Happy Farmstead Home," *Lincoln Star*, Sept. 8, 1935.
59. House of Representatives, *Select Committee of the House Committee on Agriculture to Investigate the Activities of the Farm Security Administration: Hearings*, 78th Cong., 1st sess., 1943, H. Doc. 978, 1125, 1128–29, 1130–31.
60. Resettlement Administration, Region VII, *Resettlement News*, vol. 1, no. 2 (n.d.).
61. "Hits Project at FERAville," *Omaha World-Herald*, July 27, 1935.
62. "'Deserted' FERAville Under Padlock, Study Plans for 200-Home Project," *Omaha World-Herald*, Nov. 10, 1935.
63. "Tugwell Plans to Guide Migration of Farmers from Devastated Areas," *New York Times*, July 29, 1936.
64. Lyle T. Alverson, Acting Executive Director of the National Emergency Council to President Roosevelt, Aug. 12, 1936, Folder: Drought Conditions, OF 987, FDRL.
65. "Resettlement: Its Job," *Nation's Agriculture*, Feb. 1936, 31.
66. Henry Morgenthau Jr., Diaries, April 7, 1936, FDRL.
67. The Reminiscences of Howard Tolley, Oct.–Dec. 1952, p. 343 in the Columbia University Oral History Research Office Collection.

4. "Rehabbers" in the Great Plains

1. Mrs. W. L. Hannon to Eleanor Roosevelt, El Dorado, Kans., 1939, Folder: Farm Security Administration, 1939, OF 1568 Misc., FDRL.
2. House, *Select Committee to Investigate the Farm Security Administration: Hearings*, 992.
3. Anthony J. Badger, *The New Deal: The Depression Years, 1933–1940* (New York: Noonday Press, 1989), 168–69.
4. Danbom, *Born in the Country*, 216, 219.
5. Bernard DeVoto, "The West: A Plundered Province," *Harper's Magazine*, Aug. 1934, 360, 355.
6. Leonard J. Arrington, "The New Deal in the West: A Preliminary Statistical Inquiry," *Pacific Historical Review* 38 (1969): 312–14; Leonard J. Arrington, "The Sagebrush Resurrection: The New Deal Expenditures in the Western States, 1933–1939," *Pacific Historical Review* 52 (1983): 9, 11–12.

7. Department of Agriculture, *Agricultural Statistics, 1940* (Washington DC: GPO, 1940), 630–31.

8. Henry A. Wallace, *The Price of Vision: The Diary of Henry A. Wallace, 1942–1946,* ed. John M. Blum (Boston: Houghton Mifflin, 1973), 18.

9. Reminiscences of W. W. Alexander, Aug. 4, 1952, p. 637 in the Columbia University Oral History Research Office Collection.

10. Reminiscences of W. W. Alexander, 634–35.

11. "Rural Families Promised Help from Tugwell," *Sioux Falls (S.Dak.) Argus Leader,* Aug. 28, 1935; figures from "RA Office Here, Under Trained Personnel," *Lincoln Journal,* Jan. 26, 1936.

12. "Aid for Farmers," *Topeka State Journal,* Sept. 14, 1935.

13. "Carlson Asks Relief for Farmers in the 6th," *Topeka Daily Capital,* Sept. 16, 1935.

14. "No Aid Now for Farm Families of Western Kansas," *Garden City (Kans.) Telegram,* Sept. 30, 1935.

15. Howard Wood, "Report on Rural Rehabilitation," Jan. 21, 1936, Fargo, N.Dak., Folder: Talks: J. O. Bergheim and Howard Wood, Series 1, General Correspondence (1935–1942), FHA Records, RG 96, National Archives–Central Plains Region, Kansas City, Mo. (hereafter NA–CPR).

16. *Appendix to the Congressional Record,* 76th Cong., 3d sess., 1940, 86, pt. 14:1432–33.

17. H. M. Pettygrove to Roosevelt, Apr. 15, 1936, Oxford, Nebr., File 209, Cochran Papers, NSHS.

18. L. A. White to Robert L. Cochran, May 13, 1936, Lincoln, Nebr., File 209, Cochran Papers, NSHS.

19. Rexford Tugwell to President Roosevelt, Washington DC, Sept. 24, 1935, Folder: FSA, Sept.–Nov. 1935, OF 1568, FDRL.

20. "Cal Ward Is Called to His New Field of Work," *Kansas Union Farmer,* Sept. 5, 1935.

21. "RA Office Here" *Lincoln Journal.*

22. Report, Wayne Green, Chairman of Osborne County, Folder: Reports by Campaign Chairmen on the Campaign of 1940: Kansas, DNC Papers, FDRL.

23. Report, H. H. Humphrey, Chairman of Beadle County (South Dakota), File: County Chairmen Reports for Campaign of 1940, Misc. Papers, 1932–1948, DNC Papers, FDRL.

24. "South Dakota," Folder: Complaint Reports by State, 1941, DNC Papers, FDRL.

25. Press Release, W. W. Alexander, Resettlement Administration, Jan. 22, 1937, Folder: Tenant Farmers, Subsistence Homesteads: Post Construction Progress, Committee on Public Administration, RG 200, National Archives Gift Collection, National Archives, Washington DC.

26. "Cal Ward," *Kansas Union Farmer.*

27. "Two Big Jobs Face Cal Ward on Farm Project," *Topeka Capital,* Sept. 15, 1935.

28. L. A. White, Radio Speech, Feb. 21, 1936, Folder: Talks: Nebraska, State Officials and Supervisors, Series 1: Gen. Corres. (1935–1942), FHA Records, RG 96, NA-CPR.

29. Department of Agriculture, Farm Security Administration, *Annual Report, 1941* (Washington DC: GPO, 1941), 8.

30. White, Radio Speech, NA-CPR.

31. Howard Wood, "Report on Rural Rehabilitation," FHA Records, RG 96, NA-CPR.

32. "O. Leonard Orvedal," *North Dakota History* 44 (1977): 69.

33. "Federal Relief Granted 11,000 Farm Families," *Topeka Capital,* Nov. 16, 1935; "$2,000,000 More Needed for RA-Loans—Cochran," *Lincoln Star,* Apr. 8, 1936; "S.D. to Draw More Farm Aid," *Yankton (S.Dak.) Press and Dakotan,* Nov. 19, 1935.

34. Farm Security Administration, *County Farm Security Advisory Councils and Committees Guidebook* (n.p., c. 1941), FSA Instruction 403.1 and 731.1.

35. Farm Security Administration, *County Farm Security Guidebook,* FSA Instruction 731.10.

36. Farm Security Administration, *County Farm Security Guidebook,* FSA Instruction 731.10.

37. "Minimum Satisfactory Level of Living for Families," memo, Nov. 27, 1941, Folder: Correspondence, 1937–1945, Land Tenure Studies and Conferences, Elco Greenshields Papers, HHPL.

38. Helen Buhler Ossman, Farm Security Administration, interview with author, Topeka, Kans., Oct. 7, 1996.

39. Clyde Abbott to Clifford Hope, Elkhart, Kans., Dec. 11, 1941, and Clifford Hope to Clyde Abbott, Dec. 24, 1941, Folder: Rehabilitation, Departmental Correspondence, 1941–1943, Agriculture Series, Hope Papers, KSHS.

40. Department of Agriculture, Farm Security Administration, "Progress Report of Rural Rehabilitation, Region VII: North Dakota, South Dakota, Nebraska, Kansas [1936–1937]," Jan. 1, 1938, mimeo, 4–15.

41. Nebraska, Agricultural Experiment Station, *Farm Tenancy in Box Butte County, Nebraska,* 13–15.

42. State Letter #267: "1938 Loan Program," Folder: FSA, Series 78G-4 to 78H-1, Agricultural Relief Programs, Val Kuska Papers, NSHS.

43. Farm Security Administration, "Progress Report, Region VII," 4, 8–13.

44. "Nebraska FSA to Get Funds for 39–40 Work," *Alliance (Nebr.) News,* July 20, 1939.

45. Cal Ward to Clifford Hope, Feb. 9, 1941, Folder: Rehabilitation, Hope Papers, KSHS; Clifford Hope to Alfred Crotinger, Feb. 11, 1941, Folder: Rehabilitation, Hope Papers, KSHS.

46. Farm Security Administration, "Progress Report, Region VII," 4, 8–13.

47. Department of Agriculture, *Agricultural Statistics, 1943* (Washington DC: GPO, 1943), 460.

48. Farm Security Administration, "Progress Report, Region VII," 7, 10, 13, 15.

49. Reminiscences of W. W. Alexander, 443.

50. "Living Standards of FSA Families Up," *Topeka Capital,* Dec. 30, 1938.

51. Department of Agriculture, Farm Security Administration, *Annual Report, 1945–1946,* appendix, table 1.

52. Cal Ward, "Resettlement Administration supplements activities of commercial banks . . . ," Folder: Dept. Correspondence, 1937–1938, Rehabilitation (Dept. Agriculture), General Correspondence, Hope Papers, KSHS.

53. "FSA Meeting is Over," *Salina (Kans.) Journal,* Mar. 18, 1939.

54. South Dakota, State Planning Board, *Agricultural Resources of South Dakota* (Brookings: State Planning Board, Mar. 1934), III–141.

55. Reminiscences of Howard Tolley, 478.

56. John D. Black, "Report upon the FSA," Manuscript, Folder: FSA: May 1943 Report to Chester Davis, Research Projects and Related Materials, John D. Black Papers, State Historical Society of Wisconsin (hereafter SHSW), Madison.

57. Farm Security Administration, *Annual Report, 1945–1946,* appendix, table 1.

58. Clingerman, *Field Man,* 57–60.

59. Farm Security Administration, *Annual Report, 1941,* 42.

60. File case nos. 32-14-138652, 32-60-126041, Folder: Articles and Press Releases: Case Histories, Series 1: Correspondence, 1935–1942, FHA Records, RG 96, NA-CPR. Clients' names from case files are pseudonyms.

61. Elco Greenshields, "Farm Tenure Improvement as an Aid to Conservation," Folder: Farm Tenure Improvement, Land Studies and Conferences, Greenshields Papers, HHPL.

62. Resettlement Administration, *Compensation as a Means of Improving the Farm Tenancy System,* 27, 86.

63. Department of Agriculture, Farm Security Administration, Region VII, "Improved Tenure is Part of FSA Farm Program," press release, Lincoln, Nebr., May 11, 1938, Vertical File: Farm Tenancy, NSHS.

64. "Confidential Report: To the Farm Security Administration on Tenure Improvement Work Conducted by County Supervisors in Selected Counties in Region VII," Folder: Report on Activities in the Upper Midwest, 1940, Land Tenure Studies and Conferences, Greenshields Papers, HHPL.

65. "Confidential Report on Tenure Improvement," 141, 268–69, 68–69, 108.

66. "Confidential Report on Tenure Improvement," 108, 269, 142–43.

67. "Confidential Report on Tenure Improvement," 73–74.

68. "Confidential Report on Tenure Improvement," 73–74, 145–46, 272.

69. "Confidential Report on Tenure Improvement," 85–86, 114, 157–58, 276–77.

70. James Marten, "Continuity and Change on the Twentieth-Century Farm," 46.

71. C. F. Parsons to Senator Arthur Capper, Aug. 25, 1939, Sterling, Kans., Folder: Tenancy, General Correspondence, Arthur Capper Papers, KSHS.

72. N. E. Hansen, "The Master Key to the Farm Tenant Problem," Folder: Tenant Farming, 1935–1938, OF 1650, FDRL.

73. House of Representatives, *Farm Tenancy: Hearing Before the House Committee on Agriculture,* 75th Cong., 1st sess., 284–87.

74. "Demonstration Project," editorial, *Omaha World-Herald,* Apr. 14, 1938.

75. Hansen, "Master Key to the Farm Tenant Problem."

76. Thomas A. Leadley, "Tenancy: Deep-rooted Disease," *Nebraska Farmer,* Jan. 3, 1937, 5; *Congressional Record,* 75th Cong., 1st sess., June 28, 1937, 6458.

77. Earl Singleton to Clifford Hope, Hutchinson, Kans., Apr. 4, 1937, Folder: Farm Tenant (Legis.), Hope Papers, KSHS; Dr. D. E. Maxwell to Karl Stefan, n.d., Columbus, Nebr., Folder: Farm Tenancy, Karl Stefan Papers, NSHS.

78. Anton Odvarka to President Roosevelt, Feb. 16, 1937, Clarkson, Nebr., Folder: Farm Tenancy, Stefan Papers, NSHS.

79. Reminiscences of W. W. Alexander, 586.

80. House, *Farm Tenancy,* 316.

81. "Land Loans Not the Solution," editorial, *Nebraska Union Farmer,* Mar. 22, 1943, 4.

82. Wallace, "Rural Poverty," from Rasmussen, *Agriculture in the United States: A Documentary History,* 3:2850.

83. Hope to L. R. French, n.p., Jan. 18, 1937, Folder: Farm Tenant (Legis.), Hope Papers, KSHS.

84. House, *Farm Tenancy*, 103, 105.

85. House, *Farm Tenancy*, 103, 105–6.

86. *Congressional Record*, 75th Cong., 1st sess., Aug. 17, 1937, 9166.

87. Farm Security Administration, *Annual Report, 1945–1946*, appendix, tables 2 and 3.

88. House, *Select Committee to Investigate the Farm Security Administration: Hearings*, 1015–17.

89. "Elk Point Tenant Gets $12,100 FSA Loan to Start on His Own," *Sioux Falls Argus Leader*, Jan. 27, 1940.

90. "Tenant Purchase Plan Seeks to Make Owners of Renters," *Bismarck Tribune*, Dec. 6, 1940.

91. Farm Security Administration, *Annual Report, 1941*, 31–32.

5. Farming in Place

1. Department of Agriculture, Soil Conservation Service, *Soil Survey of Barnes County, North Dakota* (n.p., June 1990), 1–2, General Soil Map.

2. Bureau of the Census, *Sixteenth Census: 1940, Agriculture*, vol. 2, : pt. 1, county table 19.

3. Thatcher to Congress, Jan. 30, 1934, Hope Papers, KSHS.

4. Works Progress Administration, *Final Statistical Report of the FERA*, 279.

5. Figures from Howard Wood, "Report on Rural Rehabilitation," FHA Records, RG 96, NA-CPR.

6. Soil Conservation Service, *Soil Survey of Coffey County, Kansas* (n.p., June 1982), 72–73.

7. Bureau of the Census, *Sixteenth Census: 1940, Agriculture*, vol. 2, pt. 1, county table 19.

8. Works Progress Administration, *Final Statistical Report of FERA*, 263.

9. Kansas, Emergency Relief Committee, *Public Welfare Service in Kansas, 1934*, Bulletin no. 289 (Nov. 1, 1935): 735.

10. Department of Agriculture, *Family Income and Expenditures*, 178–79.

11. Bureau of the Census, *Sixteenth Census: 1940, Agriculture*, vol. 2, pt. 1, county table 17.

12. Rural Rehabilitation Case #40-2-200572, Barnes County Rural Rehabilitation Loan Case Files, 1934–1944 (hereafter BCRRLCF), FHA Records, RG 96, National Archives–Rocky Mountain Region, Denver, Colo. (hereafter NA–RMR).

13. North Dakota, Agricultural Experiment Station, *Farm Tenancy and Rental Contracts in North Dakota*, Bulletin no. 289 (Fargo: North Dakota Agricultural Experiment Station, Nov. 1937), 9, 15.

14. Bureau of the Census, *Sixteenth Census: 1940, Agriculture*, vol. 2, pt. 1, county table 17.

15. Case File #18-16-13339, Coffey County Rural Rehabilitation Loan Case Files, 1934–1944 (hereafter CCRRLCF), FHA Records, RG 96, NA-CPR.

16. Rural Rehabilitation Case #40-2-200572, BCRRLCF, RG 96, NA-CPR.

17. Rural Rehabilitation Case #18-16-13339, CCRRLCF, RG 96, NA-CPR.

18. Rural Rehabilitation Case #40-2-27011, BCRRLCF, RG 96, NA-RMR.

19. Rural Rehabilitation Case #40-2-244290, BCRRLCF, RG 96, NA-RMR.

20. Rural Rehabilitation Case #40-20-218237, BCRRLCF, RG 96, NA-RMR.

21. Rural Rehabilitation Case #40-2-203984, BCRRLCF, RG 96, NA-RMR.

22. Rural Rehabilitation Case #40-2-251001, BCRRLCF, RG 96, NA-RMR.

23. Rural Rehabilitation Case #40-2-238723, BCRRLCF, RG 96, NA-RMR.

24. Rural Rehabilitation Case #40-2-272028, BCRRLCF, RG 96, NA-RMR.

25. Rural Rehabilitation Case #40-2-252687, BCRRLCF, RG 96, NA-RMR.

26. Rural Rehabilitation Case #18-16-17052, CCRRLCF, RG 96, NA-CPR.

27. Rural Rehabilitation Case #18-16-18584, CCRRLCF, RG 96, NA-CPR.

28. Rural Rehabilitation Case #18-16-219, CCRRLCF, RG 96, NA-CPR.

29. Rural Rehabilitation Case #18-16-219, CCRRLCF, RG 96, NA-CPR.

30. Rural Rehabilitation Case #18-16-16130, CCRRLCF, RG 96, NA-CPR.

31. Rural Rehabilitation Case #18-16-32837, CCRRLCF, RG 96, NA-CPR.

32. See also Rural Rehabilitation Cases #18-16-21229 and #18-16-20718, CCRRLCF, RG 96, NA-CPR.

33. Kansas, Agricultural Experiment Station, and State Planning Board, *Agricultural Resources of Kansas*, Bulletin no. 31 (Manhattan: Kansas State College, 1937), 41–43.

34. Soil Conservation Service, *Soil Survey of Coffey County, Kansas*, 29.

35. Rural Rehabilitation Case #18-16-34559, CCRRLCF, RG 96, NA-CPR.

36. Rural Rehabilitation Case #18-16-20718, CCRRLCF, RG 96, NA-CPR.

37. Rural Rehabilitation Case #18-16-39098, CCRRLCF, RG 96, NA-CPR.

38. Rural Rehabilitation Case #40-2-200572, BCRRLCF, RG 96, NA-RMR.

39. Rural Rehabilitation Case #18-16-13339, CCRRLCF, RG 96, NA-CPR.

40. Rural Rehabilitation Case #18-16-20652, CCRRLCF, RG 96, NA-CPR.

41. Rural Rehabilitation Case #18-16-219, CCRRLCF, RG 96, NA-CPR.

42. "Confidential Report on Tenure Improvement Work," 263–64, Greenshields Papers, HHPL.

43. Rural Rehabilitation Case #18-16-9836, CCRRLCF, RG 96, NA-CPR.

6. Politics, War, and the Downfall of the FSA

1. Anthony Badger, *The New Deal: The Depression Years, 1933–1940* (New York: The Noonday Press, 1989), 10, 306, 2, 167.

2. Badger, *The New Deal*, 169.

3. "Capper Lauds Roosevelt," *Wichita Beacon*, July 23, 1933.

4. South Dakota, *Legislative Manual, 1939* (n.p.), 457.

5. Works Progress Administration, *Nebraska Party Platforms, 1858–1940* (Lincoln: University of Nebraska, 1940), 498a; University of North Dakota, Bureau of Governmental Affairs, *A Compilation of North Dakota Political Party Platforms, 1884–1978* (Bismarck: North Dakota State Library, 1979), 178; South Dakota, *Legislative Manual, 1941*, 370, 366.

6. Kenneth S. Davis, *Kansas: A Bicentennial History* (New York: W. W. Norton, 1976), 178.

7. "Shall North Dakota Cooperate?" editorial, *Devils Lake (N.Dak.) Journal*, Aug. 15, 1934.

8. "Economic Crime," editorial, *Lincoln Star*, Sept. 26, 1936.

9. "The American Way," editorial, *Lincoln Star*, Sept. 28, 1936.

10. "Repaying Loans," editorial, *Lincoln Star*, Oct. 7, 1936.

11. "Farmers' Opinions on Agricultural Legislation," Folder: Name and Subject File, 1933–1936, X-Y-Z, OF 227a, FDRL.

12. "Farmers Back from Trek to Washington," *Nebraska State Journal*, May 20, 1935.

13. "Votes for Checks," *Kansas City Star*, n.d. (1935), Folder: Agricultural Clippings, Jan.–Aug. 1935, Post-Presidential Subject File, HHPL.

14. "Dakotans Starving?" letter, *Fargo Forum*, Mar. 24, 1936.

15. T. J. Edmonds to Harry Hopkins, Nov. 27, 1933, Folder: North Dakota: Field Reports, 1933–1936, Federal Relief Agency Papers, Hopkins Papers, FDRL.

16. Mrs. Leon Dallen to President Roosevelt, Miltonvale, Kans., Aug. 1, 1935, OF 227, FDRL.

17. W. D. Lynch to James Farley, La Moure, N.Dak., Sept. 15, 1936, Folder: North Dakota, Correspondence of J. A. Farley, Chairman, 1936, DNC Papers, FDRL.

18. "Republicans Labor to Find Attractive Farm Program," *Baltimore Sun*, Jan. 21, 1936.

19. John T. Flynn, "Bids for the Homestead of the Free," *Collier's*, Jan. 7, 1939, 49.

20. Catherine McNicol Stock, *Main Street in Crisis: The Great Depression and the Old Middle Class on the Northern Plains* (Chapel Hill: University of North Carolina Press, 1992), 42–43, 62, 86–87, 120.

21. James Beddow, "Midwestern Editorial Responses to the New Deal, 1932–1940," *South Dakota History* 4 (1973): 10.

22. James B. Beddow, "Depression and New Deal: Letters from the Plains," *Kansas Historical Quarterly* 43 (1977): 152.

23. "Drouth Aid," *Nebraska Farmer*, Nov. 2, 1940, 8.

24. "The State Director Speaks," editorial, *Ravenna (Nebr.) News*, Aug. 23, 1940.

25. James L. Forsythe, "Clifford Hope of Kansas: Practical Congressman and Agrarian Idealist," *Agricultural History* 51 (1977): 412–17.

26. "The Price of Parity," editorial, *Omaha World-Herald*, Oct. 22, 1937.

27. "What is 'Compulsion,'" editorial, *Omaha World-Herald*, Jan. 10, 1938.

28. Editorial, *Topeka Journal*, Dec. 21, 1938.

29. South Dakota, *Legislative Manual, 1935*, 524.

30. Works Progress Administration, *Nebraska Party Platforms, 1858–1940*, 477, 478, 480–81.

31. South Dakota, *Legislative Manual, 1941*, 369.

32. Works Progress Administration, *Nebraska Party Platforms, 1858–1940*, 494c.

33. "McKelvie Answers 'A Reader,'" editorial, *Nebraska Farmer*, Oct. 24, 1936, 6; "Where Has Your Market Gone?" political advertisement, *Nebraska Farmer*, Oct. 24, 1936, 32.

34. Jerome Tweton and Daniel F. Rylance, *The Years of Despair: North Dakota in the Great Depression* (Grand Forks, N.Dak.: Oxcart Press, 1973), 18.

35. University of North Dakota, *Compilation of North Dakota Political Party Platforms*, 168.

36. "Let's Start at Home," editorial, *Sioux Falls Argus Leader*, Oct. 29, 1941.

37. Michael V. Namorato, *Rexford G. Tugwell: A Biography* (New York: Praeger, 1988), 95.

38. Peter Fearon, *War, Prosperity and Depression: The United States Economy, 1917–1945* (Lawrence: University Press of Kansas, 1987), 266–68.

39. Bureau of the Census, *United States Census of Agriculture: 1945*, vol. 1, pts. 11–13, state tables 1–3.

40. "Agriculture Vital Phase of Defense, Says Grange Head," *Hastings (Nebr.) Daily Tribune*, Oct. 1940, File 77-A-1, Misc. Correspondence, Clippings 1925–1940, Series: Agricultural Situation, Kuska Papers, NSHS.

41. "Resolutions—Governors' Conference, March 15, 1943," Folder: Gubernatorial Correspondence and Subject, Midwest Governors' Conference, Des Moines, Iowa, 1943, Series: Gubernatorial, 1942–1945, Bourke Hickenlooper Papers, HHPL.

42. Elizabeth Chitty to President Roosevelt, Bigelow, Kans., Sept. 17, 1942, Folder: Farm Matters, 1942, OF 227, FDRL.

43. Farmer unknown. Quoted from Robert Karolevitz, "Life on the Home Front: South Dakota in World War II," *South Dakota History* 19 (1989): 411.

44. "Analysis of Letters from County Chairmen in Answer to a Form Letter Sent by the Chairman of the Democratic National Committee, Jan. 15, 1944," Folder: County Chairman's Survey, 1943, Miscellaneous Papers, DNC Papers, FDRL.

45. Halvor L. Halvorson to Frank C. Walker, Minot, N.Dak., Nov. 24, 1943, Folder: North Dakota, Series: 1943 Political Situation by States, DNC Papers, FDRL.

46. Bela Gold, *Wartime Economic Planning in Agriculture: A Study in the Allocation of Resources* (New York: Columbia University Press, 1949), 181.

47. Msgr. Luigi Ligutti, Des Moines, Iowa, Dec. 3, 1943, Series 1, National Catholic Rural Life Conference Archives, Marquette University Archives, Milwaukee.

48. Bishop Vincent J. Ryan to Usher Burdick, Bismarck, N.Dak., Dec. 31, 1941, inserted in *Appendix to the Congressional Record*, 77th Cong., 2nd sess., 88, pt. 8:A62; Bishop Frank Thill to Arthur Capper, Concordia, Kans., Feb. 26, 1943, Folder: Farm Security Administration, Agricultural Correspondence, Capper Papers, KSHS.

49. Congress of Industrial Relations, Press Release, Apr. 15, 1943, Folder: Farm Security Administration, Series: File of the Economist, American Federation of Labor Papers (hereafter AFL Papers), SHSW.

50. Ben Henry, President, Iowa-Nebraska Congress of Industrial Organizations Council, to Karl Stefan, Apr. 19, 1943, Folder 46: Farm Security Administration, Stefan Papers, NSHS.

51. Ralph Barney to Karl Stefan, Apr. 3, 1943, Kearney, Nebr., Folder 46: Farm Security Administration, Stefan Papers, NSHS.

52. J. Rex Henry to Karl Stefan, May 26, 1943, Fremont, Nebr., Folder 46: Farm Security Administration, Stefan Papers, NSHS.

53. Mrs. Wilks G. Harper to Clifford Hope, Larned, Kans., Jan. 9, 1942, Folder: Rehabilitation, Department Correspondence, 1941–1943, Agriculture, Hope Papers, KSHS.

54. Martin Jacobson to William Langer, Mohall, N.Dak., Apr. 30, 1943; Hilda and Joseph Metcalf, Fort Rice, N.Dak., Apr. 6, 1943. Both letters inserted in *Congressional Record*, 78th Cong., 1st sess., May 10, 1943, 89, pt. 3:4119.

55. "Question of Aiding or Ousting 3,000,000 Farms," *Bottineau (N.Dak.) Courant*, Feb. 4, 1943, inserted in *Congressional Record*, 78th Cong., 1st sess., Feb. 11, 1943, 89, pt.1:828.

56. Robert C. Brower to Karl Stefan, Fullerton, Nebr., June 16, 1943, and Karl Stefan to Robert C. Brower, June 19, 1943, both from Folder: Farm Security Administration, Stefan Papers, NSHS.

57. Press Release, Franklin D. Roosevelt, White House, Washington DC, July 3, 1942, OF 1568, FDRL.

58. James S. Patton, et al., to President Roosevelt, Washington DC, June 20, 1942, OF 1568, FDRL.

59. Robert L. Tontz, "Membership of General Farmers' Organizations, United States, 1874–1960," *Agricultural History* 38 (1964): 155.

60. Conrad and Conrad, *50 Years: North Dakota Farmers Union*, 67–68.

61. "Remarks of James G. Patton, President, National Farmers Union on N.B.C. Farm and Home Hour," Feb. 28, 1942, Folder: FSA 1939–1943, Gardner Jackson Papers, FDRL.

62. James G. Patton and M. W. Thatcher to Representative Jed Johnson, Washington DC, Mar. 5, 1942, inserted in *Congressional Record*, 77th Cong., 2d sess., Mar. 6, 1942, 88, pt. 2:2029.

63. Sidney Baldwin, *Poverty and Politics: The Rise and Decline of the Farm Security Administration* (Chapel Hill: University of North Carolina Press, 1968), 356.

64. National Farmers Union, Press Release, Apr. 12, 1943, Folder: Farm Security Administration, Series: Files of the Economist, AFL Papers, SHSW.

65. "Congressional 'Economy Bloc' Attacks FSA," editorial, *Kansas Union Farmer*, Jan. 21, 1943, 8.

66. "FSA—Letters Needed to Overcome Stall on Farm Home Bill," *Kansas Union Farmer*, Apr. 27, 1944, 1.

67. "Congressman Lambertson Says Farmers Should Become Hired Hands!" editorial, *Kansas Union Farmer*, May 6, 1943, 8.

68. "Want Independent Farmers," editorial, *Nebraska Union Farmer*, Apr. 26, 1944, 4.

69. Chamber of Commerce of the United States, *Rural Relief and Rehabilitation under the Farm Security Administration* (Washington DC: Chamber of Commerce, May 1942), 7, 11, 13.

70. Raymond Gilkeson to Arthur Capper, Topeka, Kans., Mar. 8, 1943, Folder: Farm Security Administration, Agricultural Correspondence, Capper Papers, KSHS.

71. House, *Select Committee to Investigate the Farm Security Administration: Hearings*, 1461–62.

72. Department of Commerce, *Statistical Abstract, 1944–1945,*, 263.

73. Letter from J. H. Beam to Edward O'Neal, Seneca, S.Dak., Feb. 9, 1942, inserted in *Congressional Record*, 77th Cong., 2d sess., May 18, 1942, 88, pt. 3:4300–4301.

74. Letter from Guy Earnest to Edward O'Neal, Ravenna, Nebr., Feb. 16, 1942, inserted in *Congressional Record*, 77th Cong., 2d sess., 4301.

75. Tontz, "Membership of General Farmers' Organizations," 156.

76. Orville M. Kile, *The Farm Bureau through Three Decades* (Baltimore: Waverly Press, 1948), 323.

77. Advertisement, *Jetmore (Kans.) Republican*, c. Dec. 1945, Folder: Hodgeman County, Farm Bureau, News Releases, 1945, Records of the Hodgeman County Extension Service: Official Files, Correspondence and Reports, 1916–1953, KSHS.

78. "Keeping Tab on Washington," *Nation's Agriculture*, May 1943, 3.

79. "Excerpt from Statement of Edward O'Neal," inserted in *Congressional Record*, 78th Cong., 2d sess,, May 18, 1942, 88, pt. 3:4291.

80. "Keeping Tab on Washington," *Nation's Agriculture*, June 1943, 4, 10–11.

Conclusion

1. "A Land Boom Looms," *Capper's Farmer*, Nov. 1943, 7.

2. "Farm Owners Foresee 1943 Boom—Nebraska Land Prices Due to Soar," *Lincoln Star*, Jan. 8, 1943.

3. "Buying North Dakota," editorial, *Williams County (N.Dak.) Press*, Apr. 8, 1943.

4. "House Concurrent Resolution 5," Folder: FSA, HR.78A-HI.7, Petitions, Papers of the 78th Congress, Committee on Agriculture, Records of the United States House of Representatives, National Archives, Washington DC.

5. Farm Security Administration, *Annual Report, 1945–1946*, appendix, table 5.

6. "Back to the Land," *Country Gentleman*, Oct. 1944, 20.

7. "Farm Opportunities for Veterans," *Nation's Agriculture*, Sept. 1944, 4, 12.

8. South Dakota, Agricultural Experiment Station, *Buying Land? Avoid Foreclosure*, Circular no. 36 (Brookings: South Dakota Agricultural Experiment Station, Jan. 1942), 3, 10–11.

9. Wes Jackson, *New Roots for Agriculture* (San Francisco: Friends of the Earth, with the Land Institute, 1980), 17.

10. Garry Wills, *A Necessary Evil: A History of American Distrust of Government* (New York: Simon and Schuster, 1999), 15, 21, 320.

11. Bill Clinton, "What Good is Government . . . ," and Newt Gingrich, "And Can We Make it Better?," *Newsweek*, Apr. 10, 1995, 20, 22, 25.

12. Department of Commerce, *Statistical Abstract, 2000* (Washington DC: GPO, 2000), 671; Department of Agriculture, *1997 Census of Agriculture, State and County Data*, vol. 1: *Geographic Area Series*, pts. 16, 27, 34, 41 (Washington DC: United States Department of Agriculture, n.d.), table 1.

13. Bureau of Agricultural Economics, *Culture of a Contemporary Rural Community*, by Earl Bell, 56.

14. Thomas D. Isern, "The American Dream: The Family Farm in Kansas," *The Midwest Quarterly* 26 (1985): 367.

SOURCES

Manuscripts
Franklin D. Roosevelt Library
Democratic National Committee Papers
Harry L. Hopkins Papers
Gardner Jackson Papers
Franklin D. Roosevelt—Papers as President

Herbert Hoover Presidential Library
Elco Greenshields Papers
Bourke B. Hickenlooper Papers
Herbert Hoover—Post-Presidential Papers
J. Franklin Thackery Papers (unprocessed)

Kansas State Historical Society–Topeka
Atchison, Topeka, & Santa Fe Railway Co. Agricultural
Development and Publicity Office Papers
Arthur Capper Papers
Hodgeman County Extension Service—Office Files
Clifford Hope Papers

Marquette University Archives–Milwaukee
National Catholic Rural Life Conference Archives

National Archives–Central Plains Region–Kansas City, Missouri
Records of the Farm Home Administration, Record Group 96:
Coffey County, Kansas Rural Rehabilitation Loan Case Files
General Correspondence File

National Archives–Rocky Mountain Region–Denver, Colorado
Records of the Farm Home Administration, Record Group 96:
Barnes County, North Dakota Rural Rehabilitation Loan Case Files

National Archives–Washington DC
National Archives Gift Collection
Records of the United States House of Representatives

Nebraska State Historical Society–Lincoln
Governor Robert Cochran Papers
Val Kuska Papers
Karl Stefan Papers

State Historical Society of Wisconsin–Madison
American Federation of Labor Papers
John D. Black Papers

University of Kansas, Kansas Collection–Lawrence
John Stutz Papers

Essential Secondary Sources

Those interested in long-term agricultural trends in the United States should see R. Douglas Hurt, *American Agriculture: A Brief History* (Ames: Iowa State University Press, 1994), and Willard W. Cochrane, *The Development of American Agriculture: A Historical Analysis* (Minneapolis: University of Minnesota Press, 1979). The history of farm tenancy has attracted considerable debate among historians. Some believe it was fundamentally undemocratic and favored the landlord; others believe it was an economic tool that potentially benefited the tenant. Paul W. Gates took the first view. See his *Landlords and Tenants on the Prairie Frontier* (Ithaca, N.Y.: Cornell University Press, 1973). Donald Winters, in his *Farmers without Farms: Agricultural Tenancy Nineteenth-Century Iowa* (Westport, Conn.: Greenwood Press, 1978), takes the other side. For debates on farm tenancy and its impact on social conservation and landownership, see William H. Harbaugh's "Twentieth-Century Farm Tenancy and Soil Conservation: Some Comparisons and Questions," *Agricultural History* 66 (1992): 95–119, and Jeremy Atack's "The Agricultural Ladder Revisited: A New Look at an Old Question with Some Data for 1860." *Agricultural History* 63 (1989): 1–25.

Farming in the Great Plains has attracted various interpretations. For nineteenth-century settlement and farming, see Gilbert Fite's *The Farmers' Frontier, 1865–1900* (Albuquerque: University of New Mexico Press, 1974). Sources on twentieth-century plains agricultural trends include Donald Worster's important *Dust Bowl: The Southern Plains in the 1930s* (New York: Oxford University Press, 1979); Mary W. M. Hargreaves, *Dry Farming in the Northern Great Plains: Years of Readjustment, 1920–1990* (Lawrence: University Press of Kansas, 1993); and Bradley Howard Baltensperger, "Farm Consolidation in the Northern and Central States of the Great Plains," *Great Plains Quarterly* 7 (1987): 256–65. Two perceptive studies are David B. Danbom, "The North Dakota Agricultural Experiment Station and the Struggle to Create a Dairy State," *Agricultural History* 63 (1989): 174–86, and Richard G. Bremer, *Agricultural Change in an Urban Age: The Loup Country of Nebraska, 1910–1970*, University of Nebraska Studies, new series no. 51, (Lincoln: University of Nebraska, June 1976).

These works were very helpful for understanding general and agricultural economic trends between 1929 and 1945: Peter Fearon, *War, Prosperity and Depression: The United States Economy, 1917–1945* (Lawrence: University of Kansas Press, 1987); Robert Higgs, "Wartime Prosperity? Reassessment of the U.S. Economy in the 1940s," *Journal of Economic History* 52 (1992): 41–60; and Lawrence A. Jones and David Durand, *Mortgage Lending Experience in Agriculture* (Princeton, N.J.: Princeton University Press for the National Bureau of Economic Research, 1954).

Farm life in the Farm Belt has been a rich topic for recent histories. Two of the best general studies are Danbom, *Born in the Country: A History of Rural America* (Baltimore: John Hopkins University Press, 1995), and Mary Neth, *Preserving the Family Farm: Women, Community, and the Foundations of Agribusiness in the Midwest, 1900–1940* (Baltimore: Johns Hopkins University Press, 1995). Published photo albums, oral histories, and memoirs are especially evocative links to the rural plains. See *Farm Town: A Memoir of the 1930's* (Brattleboro, Vt.: Stephen Greene Press, 1974) with superior photos of the Horton, Kansas, area by J. W. McManigal and edited with text and additional photos by Grant Heilman. Gerald W. Wolff and Joseph H. Cash compiled and edited transcribed oral interviews for their "South Dakotans Remember the Great Depression," *South Dakota History* 19 (1989): 224–58. Harold Bennet Clingerman left a businesslike but discerning account of central-Nebraska farm life in his *Field Man: The Chronicle of a Bank Farm Manager in the 1940s* (Ames: Iowa State University Press, 1989).

The subject of farm women is truly a growth industry in rural history. Among the most valuable are Neth (above); Deborah Fink's *Agrarian Women: Wives and Mothers in Rural Nebraska, 1880–1940* (Chapel Hill: University of North Carolina Press, 1992); and Joan M. Jensen's *With These Hands: Women Working on the Land* (New York: Feminist Press/McGraw Hill, 1981).

Readers interested in immigrant farming practices in the plains will find Brian Q. Cannon's "Immigrants in American Agriculture," *Agricultural History* 65 (1991): 17–35 a good introduction to recent interpretations of the subject. See also Baltensperger's "Agricultural Change among Nebraska Immigrants, 1880–1900," in *Ethnicity on the Great Plains, ed.* Frederick C. Luebke (Lincoln: University of Nebraska Press, 1980).

Currently, there is no up-to-date comprehensive study of the two critical farm organizations, the Farm Bureau and the Farmers Union. In the meantime for the Farm Bureau, see Christiana McFayden Campbell, *The Farm Bureau and the New Deal: A Study of the Making of National Farm Policy, 1933–40* (Urbana: University of Illinois Press, 1962), and Grant McConnell's stridently critical *The Decline of Agrarian Democracy* (Berkeley: University of California Press, 1953). John A. Crampton's *National Farmers Union: Ideology of a Pressure Group* (Lincoln: University of Nebraska Press, 1965) is the best available history of that farm group. Lowell K. Dyson's *Farmer's Organization* (New York: Greenwood Press, 1986) is an invaluable guide on the topic.

There is a treasury of fine histories of the Great Depression and the New Deal. Robert S. McElvaine's *The Great Depression: America, 1929–1941* (New York: Times Books, 1984) is a readable account of the period between Black Tuesday and Pearl Harbor. President Herbert Hoover was much maligned in and out of office for his rigid ideology, as seen in Carl Degler's "The Ordeal of Herbert Hoover," *Yale Review* 52 (1963): 563–83. For a more sympathetic treatment of Hoover, which recognizes the development of New Deal policy from the 1920s, see David E. Hamilton's *New Day to New Deal: American Farm Policy from Hoover to Roosevelt, 1928–1933* (Chapel Hill: University of North Carolina Press, 1991).

The best one-volume interpretation of the New Deal is Anthony J. Badger's *The New Deal: The Depression Years, 1933–1940* (New York: Noonday Press, 1989). Leonard J. Arrington offered two articles explaining the American West's favored status in the New Deal. They are "The New Deal in the West: A Preliminary Statistical Inquiry," *Pacific Historical Review* 38 (1969): 311–16, and "The Sagebrush Resurrection: New Deal Expenditures in the Western States, 1933–1939," *Pacific Historical Review* 52 (1983): 1–16.

Two solid histories of New Deal farm policy are Theodore Saloutos's posthumous *The American Farmer and the New Deal* (Ames: Iowa State University Press, 1982) and Richard S. Kirkendall, *Social Scientists and Farm Politics in the Age of Roosevelt* (Columbia: University of Missouri Press, 1966). A pair of authors give thoughtful criticisms of the New Deal's relief programs. In *America's Struggle against Poverty 1900–1980* (Cambridge: Harvard University Press, 1981), James Patterson looks at long-term poverty before 1929 and American's views on the poor. William Bremer shows the "cruel to be kind" nature of 1930s work relief in his "Along the 'American Way': The New Deal's Work Relief Programs for the Unemployed," *Journal of American History* 42 (1975): 636–52.

New Deal farm policy combined with social concerns to create the Resettlement Administration and the Farm Security Administration. Sidney Baldwin's *Poverty and Politics: The Rise and Decline of the Farm Security Administration* (Chapel Hill: University of North Carolina Press, 1968) remains the best overall study of these rural rehabilitation agencies. Paul K. Conkin celebrates the New Deal's controversial resettlement programs and the Brain Trusters' beneficent creativity in his *Tomorrow a New World: The New Deal Community Program* (Ithaca, N.Y.: Cornell University Press, 1959). He looks at both rural and urban projects. For a sophisticated interpretation of these programs and their critics, see Cannon's *Remaking the American Dream: New Deal Resettlement in the Mountain West* (Albuquerque: University of New Mexico Press, 1996).

Some of the most relevant political histories of the 1929–1945 period examine American politics through a cultural perspective. The following works analyze the conservatism underlying American ideology. For ideology and politics at the national level, see Melvyn Dubofsky's "Not So 'Turbulent Years': Another Look at the American 1930s," *Americkastudien/American Studies* 24 (1979): 5–20, and Mark H. Leff's "The Politics of Sacrifice on the American Home Front in World War II," *Journal of American History* 77 (1991): 1296–1318. Turning to the American West and the plains, three works are notable: Ronald L. Feinman, *Twilight of Progressivism: The Western Republic Senators and the New Deal* (Baltimore: Johns Hopkins University Press, 1981); William C. Pratt's "Radicals, Farmers, and Historians: Some Recent Scholarship about Agrarian Radicalism in the Upper Midwest," *North Dakota History* 52 (1985): 12–25; and Catherine McNicol Stock's excellent *Main Street in Crisis: The Great Depression and the Old Middle Class on the Northern Great Plains* (Chapel Hill: University of North Carolina Press, 1992). For an exhaustive bibliography of the primary and secondary sources used for this book, see my dissertation submitted to the University of Kansas in 1998.

INDEX

Agricultural Adjustment Act (1933), 71–72; and parity, 71–72

Agricultural Adjustment Act (1938), 164–65

Agricultural Adjustment Administration (AAA), 28

Agricultural Experiment Service, 73, 193

agriculture, large-scale, Plains, 2, 7, 25; and cost-price squeeze, 20; and credit system, 18; and farm land expansion, 13; productivity of, 13; and technology, 14. See also capital investment in farms; mechanization, farm

agriculture, Plains: commercial orientation of, 5–6; dairy farming, 27–28; diversified farming, 26–27, 28; during 1920s, 10–11; and frontier settlement, 10; and management strategies, 24–25; post–World War II, 204–5; success stories about, during 1930s, 25–26; and World War II, 197–200

Aid to Families with Dependent Children (AFDC), 204

Alexander, Will W., 101, 116–17, 128

Allen, Henry J., 40

American Farm Bureau Federation. See Farm Bureau

Anticorporation Farming Law (North Dakota, 1932), 55

Badger, Anthony, 99, 161–62, 201

Baldwin, Sidney, 4

Bankhead-Jones Farm Tenancy Act (1937), 56, 125, 128

Barnes County ND. See rural rehabilitation case studies

Berry, Tom, 77–78, 103, 106

Binderup, Charles, 125–26

Black, John D., 118–19

borderline farmers, plains, 3, 34–35; characteristics of farms of, 30–33; definition of, 30; and geography, 30; goals of, 33–34

Bottineau County ND, 12

Buffalo Commons, 3–4

Bulow, William, 163

Burdick, Usher, 163, 186

Bureau of Agricultural Economics, 88

Byrd, Harry F., 176, 187–88

capital investment in farms: and farm foreclosures and distress transfers, 17; and mortgage debt, 16, 17–18; in Potter County SD, 17

capitalism and farmers' values, 9

Capper, Arthur, 100, 163–64, 191

Carlson, Frank, 103

Civil Works Administration (CWA), 79

class, 7

Clingerman, Harold, 23, 120

Clinton, Bill, 203

Cochran, Robert, 93

Coffey County KS. See rural rehabilitation case studies

In the Our Sustainable Future series

www.ingramcontent.com/pod-product-compliance
Lightning Source LLC
Chambersburg PA
CBHW071853270326
41929CB00013B/2209